CSR, Sustainability, Ethics & Governance

In recent years the discussion concerning the relation between business and society has made immense strides. This has in turn led to a broad academic and practical discussion on innovative management concepts, such as Corporate Social Responsibility, Corporate Governance and Sustainability Management. This series offers a comprehensive overview of the latest theoretical and empirical research and provides sound concepts for sustainable business strategies. In order to do so, it combines the insights of leading researchers and thinkers in the fields of management theory and the social sciences – and from all over the world, thus contributing to the interdisciplinary and intercultural discussion on the role of business in society. The underlying intention of this series is to help solve the world's most challenging problems by developing new management concepts that create value for business and society alike. In order to support those managers, researchers and students who are pursuing sustainable business approaches for our common future, the series offers them access to cutting-edge management approaches.

CSR, Sustainability, Ethics & Governance is accepted by the Norwegian Register for Scientific Journals, Series and Publishers, maintained and operated by the Norwegian Social Science Data Services (NSD)

Uzoechi Nwagbara • Samuel O. Idowu •
Yahaya Alhassan
Editors

Corporate Social Responsibility Disclosure in Developing and Emerging Economies

Institutional, Governance and Regulatory Issues

Springer

Editors
Uzoechi Nwagbara
University of Sunderland in London
London, UK

Samuel O. Idowu
London Metropolitan University
London, UK

Yahaya Alhassan
University of Sunderland in London
London, UK

ISSN 2196-7075 ISSN 2196-7083 (electronic)
CSR, Sustainability, Ethics & Governance
ISBN 978-3-031-61975-5 ISBN 978-3-031-61976-2 (eBook)
https://doi.org/10.1007/978-3-031-61976-2

This Springer imprint is published by the registered company Springer Nature Switzerland AG
The registered company address is: Gewerbestrasse 11, 6330 Cham, Switzerland

This book is dedicated to the cause of regulating and governing corporate social responsibility as well as institutionalizing frameworks for good corporate citizenship and business responsibility, transparency, and accountability in developing/emerging economies leading to a more ethical organizational practices.

Foreword

In today's interconnected society, the business environment has undergone significant changes. Corporate malpractices have surged in the relentless pursuit of profit and dominance, raising concerns and amplifying demands for companies to prioritize responsible behaviour. This call for a complete redefinition of the corporate ethos goes beyond the regulatory framework, as the business environment needs to re-evaluate the corporate values and principles. It is not just a matter of complying with guidelines but rather an intrinsic adoption of ethical practices prioritizing sustainability and fairness in all facets of operations, where the pursuit of profit must be reconciled with societal well-being and environmental stewardship.

In light of this grim reality, the editors of this book highlight corporate social responsibility concerns while enhancing the understanding of institutional, regulatory, and governance issues to address the 'race to the bottom' proposition. This book provides a holistic framework that encompasses corporate social responsibility-related issues by examining the interplay between legal mandates, practices, and the resultant implications for reporting standards. Moreover, it recognizes the impact of appropriate disclosure practices on sustainable development, which represents an essential issue for developing and emerging economies as the business environment is constantly evolving and stakeholders expect companies to operate responsibly and sustainably. In addition, the book provides valuable theoretical and practical implications by including insights into SMEs' responsible entrepreneurship and corporate social responsibility practices and barriers. The book's applicability in terms of practical actions provides guidance for local and multinational business organizations operating in developing and emerging economies while offering valuable insights for regulators, enthusiasts, and scholars.

The authors are great personalities, with the vast majority either operating and consulting for businesses or teaching business-related courses in corporate social responsibility. Given their expertise and experience, this book lays the foundation for appropriate disclosure practices for a more ethical, accountable, and transparent business environment.

Another notable contribution of this book is that, unlike other publications in the genre, it covers a wide range of corporate social responsibility issues across industries and regions, providing readers with in-depth knowledge of diverse but interrelated topics such as the regulatory landscape, institutional dynamics, and the profound impact of corporate social responsibility on sustainable development. This book serves as a guide to navigate the complex web of governance, societal impact, and the evolving role of businesses in addressing socio-economic challenges, presenting a compelling narrative on the ever-evolving landscape of corporate social responsibility.

This book offers a comprehensive guide for navigating the complex landscape of responsible corporate behaviour, making it an invaluable resource for professionals, academics, and enthusiasts seeking to understand and implement corporate social responsibility strategies in today's complex business environment.

Faculty of Economics
University of Alba Iulia
Alba Iulia, Romania

Dan Topor

Preface

Corporate malfeasance, irresponsibility, and unethical practices as well as the negative consequences of globalization including environmental crisis, rising income disparities, socio-economic malaise, political conflict, rising social inequalities, and the outsourcing of increasingly skilled operations to emerging and/or developing countries have triggered clamour for protection against the 'pathological pursuit of profit and power' by global corporations. This has also fuelled interest in curbing the 'anarchy of unregulated market forces' and effort to arrest corporations' 'exploitative psychopathic personalities' for them to take responsibility for their impact on society, footprint on the environment, and treacherous roles in developing countries, which are in the shadow of their influence, power, and actions.

Guided by this ratiocination, *Corporate Social Responsibility Disclosure in Developing and Emerging Economies: Institutional, Governance and Regulatory Issues* argues that understanding the institutional, regulatory, and governance issues underpinning poor CSR practice and reporting can be one of the ways to not only explicate the ravaging 'race to the bottom' thesis by multinational corporations, but to uncover and reposition the debate on corporate greed, profiteering, business irresponsibility, and lack of accountability and transparency in the methods of disclosing CSR. The notion that businesses have responsibilities to society and the environment is not new; a lot of scholarship from the Western and non-Western standpoint has addressed this issue severally and in different contexts. However, there is paucity of research on the role of corporate governance, institutional frameworks, and legal systems that frame CSR reporting and practices in developing/emerging economies as well as the legal, conventional, and taken-for-granted implications of these practices to wider stakeholders. This is the mainstay of this book!

Notwithstanding the acceptance of a context-specific significance to CSR (reporting), there is a privileging of western-centric inquiry essentially focussed on developed and progressively transitional economies underplaying and undercutting research in emerging economies. Thus, research centring on emerging/developing countries remains quite scarce with few exceptions. Given the disparities between developed and developing countries' CSR research and experiences framed by cultural, socio-economic, and political configurations, the book makes a bold effort to supplement and further extend insights into an under-researched space—developing and emerging economies.

There is no gainsaying the fact that developing countries' peculiarities about CSR (disclosure) policy regulation, corporate governance, and business responsibility continue to legitimize and reproduce norms and values, which are not in the interests of developing, marginalized economies. Thus, in rethinking scholarship from the perspective of emerging economies for a more ethical, accountable, and transparent CSR policies, actions, and activities, we must first of all explicate the institutional and regulatory environment and determinants that are shaping CSR practices and fuelling what Tom Mueller has described as 'crisis of conscience' negating an understanding of what is accepted or unacceptable practice by the corporations. A contextually relevant CSR practice and policies viewed from an exploration of the institutional, regulatory, and governance mechanisms that drive such issues is crucially important to reinvent CSR agenda, rekindle interest in sustainable development goals (SDGs), and redraw the map of corporate-stakeholder interest that are in a comatose state.

London, UK — Uzoechi Nwagbara
London, UK — Samuel O. Idowu
London, UK — Yahaya Alhassan
March 2024

Acknowledgements

We wish to express our heartfelt gratitude to all contributors who have supported this book project; they are amazing and have made praiseworthy efforts! Hand on heart, without their dedication, commitment, and hard work, this book will not have been possible. We thus sincerely express our appreciation and gratitude to all these great individuals who have contributed to the success of this book. The three editors are grateful to those of you who have stood by us with their impressive chapters in this book. Many of them despite their busy schedules felt obliged to help in putting together this very fascinating addition to the literature on how global corporations' 'race to the bottom' in developing/emerging economies because of lax regulatory systems, poor institutional frameworks, and inept corporate governance mechanisms that trigger and sustain poor CSR practices and reporting.

We would like to thank our publishing team at Springer headed by the Executive Editor, Christian Rauscher, Barbara Bethke, and other members of the publishing team who have supported this project and other projects.

We thus dedicate the book to all those who suffer from corporate greed, profiteering and crisis of conscience stemming from unregulated CSR reporting pattern, unethical behaviours of corporations, and irresponsible actions of these global firms, who operate in environments that fuel inept corporate governance and unacceptable reporting systems.

We apologize for any errors or omissions that may appear anywhere in the book, please be assured that no harm was intended to anybody. Finally, our profound gratitude goes to our families for the time granted us to write this book!

Uzoechi Nwagbara
Samuel O. Idowu
Yahaya Alhassan
March 2023

Corporate Social Responsibility Disclosure in Developing and Emerging Economies: Institutional, Governance and Regulatory Issues: An Introduction

Corporate malfeasance, irresponsibility, and unethical practices as well as the negative consequences of globalization including environmental crisis, rising income disparities, socio-economic malaise, political conflict, rising social inequalities, and the outsourcing of increasingly skilled operations to emerging and/or developing countries have triggered clamour for protection against the 'pathological pursuit of profit and power' by global corporations. This has also fuelled interest in curbing the 'anarchy of unregulated market forces' and effort to arrest corporations' 'exploitative psychopathic personalities' for them to take responsibility for their impact on society, footprint on the environment, and treacherous roles in developing countries, which are in the shadow of their influence, power, and actions.

Guided by this ratiocination, Corporate Social Responsibility Disclosure in Developing and Emerging Economies: Institutional, Governance, and Regulatory Issues argues that understanding the institutional, regulatory, and governance issues underpinning poor CSR practice and reporting can be one of the ways to not only explicate the ravaging 'race to the bottom' thesis by multinational corporations, but to uncover and reposition the debate on corporate greed, profiteering, business irresponsibility, and lack of accountability and transparency in the methods of disclosing CSR. The notion that businesses have responsibilities to society and the environment is not new; a lot of scholarship from the Western and non-Western standpoint has addressed this issue severally and in different contexts. However, there is paucity of research on the role of corporate governance, institutional frameworks, and legal systems that frame CSR reporting and practices in developing/emerging economies as well as the legal, conventional, and taken-for-granted implications of these practices to wider stakeholders. This is the mainstay of this book!

Notwithstanding the acceptance of a context-specific significance to CSR (reporting), there is a privileging of western-centric inquiry essentially focussed on developed and progressively transitional economies underplaying and undercutting research in emerging economies. Thus, research centring on emerging/developing countries remains quite scarce with few exceptions. Given the disparities between developed and developing countries' CSR research and experiences framed by cultural, socio-economic, and political configurations, the book makes a bold effort to supplement and further extend insights into an under-researched space—developing and emerging economies.

There is no gainsaying the fact that developing countries' peculiarities about CSR (disclosure) policy regulation, corporate governance, and business responsibility continue to legitimize and reproduce norms and values, which are not in the interests of developing, marginalized economies. Thus, in rethinking scholarship from the perspective of emerging economies for a more ethical, accountable, and transparent CSR policies, actions, and activities, we must first of all explicate the institutional and regulatory environment and determinants that are shaping CSR practices and fuelling what Tom Mueller has described as 'crisis of conscience' negating an understanding of what is accepted or unacceptable practice by the corporations. A contextually relevant CSR practice and policies viewed from an exploration of the institutional, regulatory, and governance mechanisms that drive such issues is crucially important to reinvent CSR agenda, rekindle interest in sustainable development goals (SDGs) and redraw the map of corporate-stakeholder interest that are in a comatose state.

The book was able to generate interests from a good number of contributors on emerging economies from four continents—Africa, Europe, North America, and South America. These countries were able to explore the theme of the book in fifteen chapters that make up the book. This explains why the book has been divided into three parts and fifteen chapters.

Part I explores critically CSR from regulatory and institutional perspectives in 5 chapters. Part II looks at CSR in terms of Governance and Sustainable Development also in 5 chapters. Part III, the final part of the book, looks at CSR and Small and Medium sized Enterprises (SME) in 2 chapters. The remainder of this introductory chapter looks at what each of the participating countries says about how the theme of the book is espoused in each of the countries where the authors are based.

The very first chapter by Osamuyimen, Uchechukwu Nwoke, and Godswill Agbaitoto takes a legal perspective to explore corporate social responsibility practices in the controversial Nigerian oil and gas sector. It does this essentially by pointing out new legal direction for a reinvented Nigeria as well as the implications for reporting CSR. The chapter is quite perspicacious as it throws up how the law, which is often flouted by multinational corporations, and related frameworks, can be instrumental in changing the game in this industry, which is sated with corruption, corporate malfeasance, stakeholder disengagement, and illegality since oil was discovered in Nigeria for several decades.

The second chapter by David Katamba, Christopher M. J. Wickert, and Cedric M. Nkiko, whose expertise and specialties are grounded in CSR and organizational behaviour, takes a humanist approach in shedding light on corporate responsibility. Central to the thesis projected in this chapter is that a humanist perspective to CSR is integral to not only reinventing the phenomenon from the developing countries' perspective, but essential for giving corporate actions and policies a human face. It further argued that a lack of humanism in practising CSR and related issues in organization could jeopardize the 'win-win' mantra in corporate-stakeholder relations (in Uganda).

In taking the debate forward, in chapter three, these contributors—Uzoechi Nwgabara, Yahaya Alhassan, Ngozi Ibeawuchi, and Jacyntha Stewart—added another dimension to the second chapter. In chapter three, they critically examined the motivations for CSR disclosure in developing country like Nigeria as well as who matters to the form, which leads to them conceptualizing normative practice, as a way of navigating stakeholders' suspicion of corporate effort at CSR, unethical practice and corporate-stakeholder dilemma as well as how it is being reported. Therefore, such re-conceptualization of CSR driver and motivation lends itself to normative organizational practice as a *sine qua non* for circumventing continual conflict, tension, and accusations of lack of transparency, accountability, and responsibility in the war CSR is operationalized by corporations.

The last but one chapter in Part I (chapter four) written by Chigozie A. Ugwoji takes an interesting and insightful approach to understanding how transformational leadership style is implicated in changing the corporate governance landscape in Nigeria, which the contributor argues could resonate with a transformed corporate governance practice, development, and 'win-win' situation for all relevant stakeholders in Nigeria. The context of Nigeria is not only relevant in shedding light on an under-researched area but contemporaneous with the current incidence of corruption, illegality, and corporate malfeasance in the nation that redouble the thesis of 'paradox of plenty'.

The final chapter in Part I—chapter five—is well crafted by Boris Christian Herbas-Torrico, Carlos Alejandro Arandia-Tavera, and Pamela Mirtha Zurita-Lara, who are seasoned scholars in their own right. Using the context of Bolivia, these contributors bring a fresh perspective to the vista of CSR reporting leveraging the customer dimension, which is rarely discussed in prior, relevant literature. The chapter points out that building trust, garnering social capital and ethical practice in the eyes of wider stakeholders including customers resonates with good corporate practice and legitimate CSR policies and reporting.

The first chapter in PART II—chapter six—is excellently written by Silvia Puiu. The overall argument in this penetrating chapter is that CSR can serve as a solution to socio-economic challenges in Romania. This chapter illuminates that argumentation and thrust using the lens of CSR in an emerging country like Romania contending how it could help in advancing the UN Sustainable Development Goals (SDGs) as well as redrawing the map of poverty alleviation, corporate wealth creation and national bliss, which are central to the tenet of SDGs. The chapter is quite topical and helps to shed light on the processes, mechanisms, and systems to make SDGs a reality in our ever unequal world, which begs the question, are corporations truly sustainable in their practices and policies?

In a similar vein, Afzal Sayed Munna, Md Sadeque Iman Shaikh, and Uzoechi Nwagbara brilliantly interrogate corporate social responsibly activities in organizations by taking a specific look at the social, economic, and environmental in chapter seven—which helps to further the discourse on SDGs. However, while chapter six investigates this phenomenon from the Romanian context, these contributors take a theoretically grounded approach, which extends the confines of what is already known about sustainable development mantra and conceptualization. Again, this chapter is an addendum to the mechanisms, theories, and steps towards realizing the ideals of SDGs for a more equal world that is hitherto ravaged by corporate greed, 'business-case' CSR and 'business-and-society' rhetoric, diminishing effort at 'business-in-society' thesis, wholesome corporate-stakeholder relations and 'win-win' logic of business-society nexus.

In chapter eight, Ursula A. Vavrik's chapter—'Advancing governance: towards sustainable development based on multilateralism and democracy: The example of Chile'—thoroughly and interesting delves into this debate of advancing the ideals of sustainable development and national prosperity as well as global business reinvention for effective governance through the lens of multilateralism and democracy, which is cardinal in corporate-stakeholder relations. The Chilean context once again throws up some of the challenges of realizing the ideals of SDGs in emerging economies, which are still in the shadow of poor political leadership, institutional voids, and corrupt practice that militate against this global clarion call for sustainable development. As can be gleaned from this chapter, democratizing discourse and the nature of relationship between relevant stakeholders in Chile as well as an alliance of multiple groups (and/or countries) in pursuing a common goal can reshape global interest in SDGs, for national and global prosperity. The novelty of this chapter resides in the concepts of democracy and multilateralism, which are often not thoroughly talked about.

In chapter nine, Henry Yamba Kamara, Uzoechi Nwagbara, Ngozi Ibeawuchi, and Chigozie A. Ugwoji use the Sierra Leonean context to further advance debate on the concept of sustainable development. The linchpin of the chapter is that CSR precedes sustainable development and one of the most effective ways to gauge ethical, legitimate, and sustainable organizational practice is how it can drive the ideals of SDGs. Exploring the nexus between CSR and sustainable development in Sierra Leone is not only important; it helps to add to the sparse literature on this country. The country has been in the jackboot of militarized environment, poor corporate-stakeholder conundrum, and MNCs' pulverizing presence in a country that essentially relies on foreign aids; yet, it is highly endowed with natural and human resources—a gripping dramatization of 'paradox of plenty' and 'resource curse'.

The final chapter in this Part—chapter ten—dwells on 'An analysis of corporate social responsibility (CSR) related objectives enshrined in the *National Oil and Gas Policy for Uganda, 2008*: A policy perspective paper'. The contributors have insightfully examined the CSR and related objectives contained in Uganda's 'National Oil and Gas Policy for Uganda, 2008', which rekindles interest in sustainable development to create awareness on Government's strategies and plans to take forward the development of the country's oil and gas sector in an effective and efficient manner for the benefit of all stakeholders specifically Ugandans. Such exploration is a welcome development in advancing the global call for countries in developing and emerging countries including the West to embrace the ideals of sustainable development for global prosperity and national bliss.

Part III starts with a very well-researched and excellent contribution in chapter eleven by John O. Okpara and Jet Mboga. Similar to the insights elaborated in chapter three about the motivations for CSR, these contributors used the context of sub-Saharan Africa's SMEs—thereby narrowing emphasis, focus, and investigation for depth. Despite these motivations, they also empirically investigated the barriers to implementing CSR by SMEs, perceptions of stakeholders about their practices and motivations. The empirical insights mediated by the research enrich the theoretical analyses in this field and further bring to bear suggestions for practical steps to navigate these barriers by SMEs for a reinvented CSR practice and sustainable, stakeholder-centric implementation of policies.

The final chapter—chapter twelve—further contextualizes Nigeria. It is well articulated and written by Adebimpe Adesua Lincoln, Brendhain Diamond, and Jane Croad leveraging the SMEs lens as does the penultimate chapter in this volume. Arguably, the overriding effort in this chapter is the examination and resonance of the concept of responsible entrepreneurship, which not only broadens the literature on SMEs but adds to the localized studies on CSR for a more nuanced, regional understanding of the phenomenon, whose operationalizing is often generalized and conflated with Western context, downplaying development and advancement of 'voices from the fringe', which are needed for a holistic approach to dealing with pressures of corporate-stakeholder conflict institutionally, nationally, and globally.

Uzoechi.nwagbara@sunderland.ac.uk Uzoechi Nwagbara
s.idowu@londonmet.ac.uk Samuel O. Idowu
yahaya.alhassan@sunderland.ac.uk Yahaya Alhassan

Contents

Part I Corporate Social Responsibility: Regulatory and Institutional Issues

1 **Corporate Social Responsibility Practices in the Nigerian Oil Industry: New Legal Direction and the Implications for Reporting** 3
Osamuyimen Egbon, Uchechukwu Nwoke, and Godswill Agbaitoro

2 **Humanism as an Institutional Driver of Corporate Social Responsibility (CSR) Activities** 21
David Katamba, Christopher M. J. Wickert, and Cedric M. Nkiko

3 **CSR Disclosure, Motivation for Disclosure, and Who Matters to the Firm: Prospecting for Normative Oraganisational Practice** 45
Uzoechi Nwagbara, Yahaya Alhassan, Ngozi Ibeawuchi, and Jacyntha Stewart

4 **The Role of Transformational Leadership in Promoting Corporate Governance Effectiveness in Nigeria** 59
Chigozie A. Ugwoji

5 **Customer Social Responsibility Reporting in Bolivia: Practices and Trends** 73
Boris Christian Herbas-Torrico, Carlos Alejandro Arandia-Tavera, and Pamela Mirtha Zurita-Lara

Part II Corporate Social Responsibility Governance and Sustainable Development

6 **CSR as a Solution for Tackling Socio-Economic Challenges in Romania** 87
Silvia Puiu

7 Unpacking Corporate Social Responsibility Activities in Organisations by Considering Social, Economic, and Environmental Challenges: A Conceptual Approach 103
Afzal Sayed Munna, Md Sadeque Imam Shaikh, and Uzoechi Nwagbara

8 Advancing Governance Toward Sustainable Development Based on Multilateralism and Democracy: The Example of Chile 125
Ursula A. Vavrik

9 Corporate Social Responsibility (CSR) and Sustainable Development: Exploring the Nexus in the Sierra Leonean Context 139
Henry Yamba Kamara, Uzoechi Nwagbara, Chigozie A. Ugwoji, and Ngozi Ibeawuchi

10 An Analysis of Corporate Social Responsibility (CSR)-Related Objectives Enshrined in the "National Oil and Gas Policy for Uganda, 2008": A Policy Perspective Paper 171
David Katamba, Bosco Amerit, Maureen Basuuta, Swithern B. Tumwine, and Ngoma Muhammed

Part III Corporate Social Responsibility and SMEs

11 CSR in SMEs in a Sub-Saharan Africa: Motivations, Perceptions, Practices, and Barriers to Implementation 193
John O. Okpara and Jet Mboga

12 Responsible Entrepreneurship: Raising Corporate Social Responsibility Awareness Among Male and Female Owned and Managed Small and Medium-sized Enterprises (SME's) in Abeokuta, Ogun State, Nigeria 209
Adebimpe Adesua Lincoln, Brendhain Diamond, and Jane Croad

Index 239

About the Editors

Uzoechi Nwagbara is a professor of management at ICT University. He has BA and MA in English as well as MSc and PhD in Human Resource Management and Management, respectively, from the University of Wales, and MEd in Higher Education as well as PG Cert in Academic Practice from the University of Sunderland, UK. He is a Fellow of the Higher Education Academy (FHEA); Member Chartered Management Institute (MCMI); Member Institute of Leadership & Management (MInstLM); and Member Community Development Exchange (MCDE); and also affiliated with British Academy of Management (BAM), European Association for the Study of Literature, Culture, and the Environment (EASLCE), Postcolonial Studies Association (PSA), and Royal African Society (RAS). Dr Nwagbara has over 35 years of teaching experience and has taught in higher education for upwards of 25 years. He is external examiner for University of Cumbria and University of Bolton. He has supervised over a dozen doctoral students to completion and published well over 100 publications in high impact journals in management, leadership, CSR, sustainability, employment relations, and human resource management. Dr Nwagbara has examined PhD students internally and externally in various universities including University of Bedfordshire, Cardiff Metropolitan University, and University of the West of Scotland amongst others. He holds Visiting Teaching Positions at the University of Sunderland in London, Coventry University London, London School of Commerce, University of the West of Scotland, The University of Law UK, University of Worcester, University of Wales Trinity Saint David, and University of Plymouth as well as Visiting Professorship (management), Coal City University. He is on the editorial boards of many peer-reviewed, international journals including *Economic Insights—Trends & Challenges*, *Journal of Pan African Studies* and *Journal of Sustainable Development in Africa*.

Samuel O. Idowu is a senior lecturer in Accounting and Corporate Social Responsibility at the Guildhall School of Business and Law, London Metropolitan University where he is currently the course leader for the MSc Corporate Social Responsibility & Sustainability and Advanced Diploma in Professional Development (ADPD) Corporate Social Responsibility & Sustainability. Samuel is a Professor of

CSR and Sustainability at Nanjing University of Finance & Economics, China. He is a fellow member of the Chartered Governance Institute, a fellow of the Royal Society of Arts, a Liveryman of the Worshipful Company of Chartered Governance Institute and a named freeman of the City of London. He is the Deputy CEO and Vice President of the Global Corporate Governance Institute, an international network of CSR scholars. Samuel has published over fifty articles in both professional and academic journals and contributed chapters in several edited books and is the Editor-in-Chief of three major global reference books by Springer—the *Encyclopedia of Corporate Social Responsibility* (ECSR), the *Dictionary of Corporate Social Responsibility* (DCSR), and the *Encyclopaedia of Sustainable Management* (ESM) and he is a Series Editor for Springer's CSR, Sustainability, Ethics, and Governance books. Dr Idowu is an Editor-in-Chief of the *International Journal of Corporate Social Responsibility* and the *American Journal of Economics and Business Administration*. He has been in academia for more than 30 years winning one of the Highly Commended Awards of Emerald Literati Network Awards for Excellence in 2008 and 2014. In 2010, one of his edited books was placed in the 18th position out of forty top Sustainability books by Cambridge University Programme for Sustainability Leadership and in 2016 one of his books won the outstanding Business Reference Book of the year of the American Library Association. In 2018 he won a CSR Leadership Award in Cologne, Germany and in 2019 he won the 101 Most Impactful CSR Leaders Award in Mumbai, India. Dr Idowu is on the Editorial Advisory Boards of the *International Journal of Business Administration and Amfiteatru Economic Journal*. He has been researching in the field of CSR since 1983 and has attended and presented papers at several national and international conferences and workshops on CSR. He has made a number of keynote speeches at international conferences and workshops and written the *foreword* to a number of leading books in the field of CSR and Sustainable Development. And he has examined on a few PhD theses in the UK, Australia, South Africa, the Netherlands, and New Zealand.

Yahaya Alhassan is a Principal Lecturer in Finance and Strategic Management at the University of Sunderland in London. He is also the Head of Academic Operations responsible for quality assurance and new programme development. He holds PhD in Business Management from Anglia Ruskin University, MBA in Finance as well as MA in Education from the University of Sunderland. He also holds BA in Business Management and a Senior Fellow of the Higher Education Academy (HEA). He has over 20 years' teaching and management experience developed during roles in the higher education sector and the consultancy industry. Dr Alhassan is a researcher, author, finance and strategic management lecturer, and consultant and has published over seventeen publications including books, book reviews, journal articles, and book chapters. Some of Dr Alhassan's publications have appeared in the UK ABS 3*** rated Journals. Dr Alhassan regularly presents papers at conferences and has presented over ten papers at various conferences since 2019. His main research interests are around development finance, microfinance, entrepreneurial finance, ethnic minority entrepreneurship and microbusiness development. His current research interests focus on the nexus between financial technology and sustainable entrepreneurship, investment in artificial intelligence, and firm growth. He has also provided consultancy services to organizations nationally and internationally that positively influenced policies, practices, and reached audiences beyond academia.

Part I
Corporate Social Responsibility: Regulatory and Institutional Issues

Corporate Social Responsibility Practices in the Nigerian Oil Industry: New Legal Direction and the Implications for Reporting
Osamuyimen Egbon, Uchechukwu Nwoke and Godswill Agbaitoro

Humanism as an Institutional Driver of Corporate Social Responsibility (CSR) Activities
David Katamba, Christopher M. J. Wickert and Cedric M. Nkiko

CSR Disclosure, Motivation for Disclosure and Who Matters to the Firm: Prospecting for Normative Organisational Practice
Uzoechi Nwagbara, Yahaya Alhassan, Ngozi Ibeawuchi, and Jacyntha Stewart

The Role of Transformational Leadership in Promoting Corporate Governance Effectiveness in Nigeria
Chigozie A. Ugwoji

Customer Social Responsibility Reporting in Bolivia: Practices and Trends
Boris Christian Herbas-Torrico, Carlos Alejandro Arandia-Tavera, and Pamela Mirtha Zurita-Lara

Chapter 1
Corporate Social Responsibility Practices in the Nigerian Oil Industry: New Legal Direction and the Implications for Reporting

Osamuyimen Egbon, Uchechukwu Nwoke, and Godswill Agbaitoro

Introduction

Over the recent decades, corporations have wielded an increasing influence on virtually every facet of the global society. This has provoked wider stakeholders' demand for corporate social responsibility (CSR) commensurate with corporate influence and power (Gray, 2006; Egbon et al., 2018). Scholars have similarly regarded corporations, especially multinational corporations (MNCs), as political actors because of their progressive venture into political spaces traditionally exclusive to the government (Frynas & Stephens, 2015; Hussain & Moriarty, 2018; Scherer & Palazzo, 2011). In developing countries such as Nigeria, the attribution of political status to MNCs especially in the oil industry is likely connected to the MNCs–government joint venture relations coupled with alleged corporate capture of the regulatory space (Noah et al., 2021). Thus, it is no surprise that the local communities expect those MNCs to undertake state-like functions (Idemudia, 2014a). Apparently, the corporations have been at the forefront of many CSR initiatives in Niger Delta communities and Nigeria-wide.

Notably, globalisation and other motivations have continued to induce the rise of CSR and the need for corporations to extend their responsibility beyond investors' interests (Ite, 2004; Scherer et al., 2009). Hence, both the conceptions and practices of CSR have characteristically been evolving in the dynamic global society (Matten & Moon, 2020). CSR practices by oil corporations in Nigeria have equally

O. Egbon (✉)
Essex Business School, University of Essex, Wivenhoe Park, Colchester, United Kingdom
e-mail: o.egbon@essex.ac.uk

U. Nwoke · G. Agbaitoro
Essex Law School, University of Essex, Wivenhoe Park, Colchester, United Kingdom
e-mail: u.nwoke@essex.ac.uk; g.agbaitoro@essex.ac.uk

U. Nwagbara et al. (eds.), *Corporate Social Responsibility Disclosure in Developing and Emerging Economies*, CSR, Sustainability, Ethics & Governance, https://doi.org/10.1007/978-3-031-61976-2_1

experienced an evolution. For example, the corporations have continued to develop, or expressed commitment to, voluntary CSR practices that have evolved from community assistance to global memorandum of understanding (GMOU[1]) in the Niger Delta where they operate (Ite, 2007; Muthuri et al., 2012). While the MNCs widely disclose those voluntary CSR practices, the extent those CSR practices have engendered sustainability-driven community development (SCD) is doubtful (Frynas, 2005) as they are apparently enlisted to procure the social license-to-operate—SLO (Idemudia & Osayande, 2018).

In enlisting CSR to obtain and sustain a SLO in the Niger Delta, the oil MNCs widely disclose those CSR initiatives via sustainability reports and other media. Idemudia and Ite (2010) noted that those CSR practices and the attendant reporting have continued to attract research attention. But such CSR research has been historically shaped around CSR practices in that it has largely focused on CSR outcome at the expense of the essential CSR process. CSR outcome, according to Idemudia and Ite (2010), includes CSR practices aimed at promoting the eradication of poverty, human rights violations, and environmental degradation. In contrast, the CSR process focuses on how the corporations conceive of and design their CSR policies which incorporate issues, namely, corporate accountability and transparency, plus the inclusive participation of the beneficiaries when designing and implementing CSR initiatives (Egbon et al., 2018; Idemudia & Osayande, 2018). Arguably, CSR practices in the Niger Delta are inadequate if decoupled from corporate environmental impacts on the communities. As such, this chapter discusses the extant voluntary CSR practices in relation to communities and the newly mandated or law-backed CSR model and reflects on the extent they promote a sustainable community development which not only addresses the socio-economic needs of the communities but also the integrity of their environment.

The remainder of this chapter is organised as follows. The next section on "CSR in the Nigerian Oil Industry" discusses the extant CSR practices of oil corporations in the Niger Delta in Nigeria, followed by a "Critique of Prior CSR Practices: Positive Affirmative Duties/Negative Injunction Duties" which reflects on those practices through the lens of positive affirmative duties and negative injunction duties. Next, section on "Institutionalised CSR Legal Framework" discusses the new CSR model enacted by the Petroleum Industry Act 2021 and the section on "Legal Approach to CSR and Its Implications for Corporate Accountability and Reporting" reflects on its implications for reporting and sustainable communities. Finally, the last section offers some concluding remarks.

[1] GMOU is supposed to reflect the ideals of sustainable development (see Chevron Nigeria, 2012), that is, the development of the present that does not undermine the interest of the present and the future generations.

CSR in the Nigerian Oil Industry

Arguably, CSR projects in the Niger Delta started in the 1960s and 1970s when the first corporations began oil exploration and production (Aaron, 2010; Idemudia, 2009, 2010a). Corporations began by making philanthropic gestures—cash payments, award of scholarships, construction of classroom blocks, roads, hospitals, and the supply of agricultural equipment. The first wave of such projects seems to have been motivated largely by business considerations and sought to do little more than satisfy 'the most-minimal' moral obligations (Nwoke, 2016). Since then, these corporations have adopted various approaches towards CSR. These include community assistance and development projects (philanthropy) and, more recently, partnership approaches. The dominant partnership approaches include government–business partnerships, business–non-governmental organisation (NGO) partnerships, and corporate–community partnerships facilitated by NGOs (Aaron, 2010; Idemudia, 2009, 2010b).

Thus, CSR initiatives have been on the rise in Nigeria with more visible CSR initiatives and strategies among MNCs in the oil industry (Frynas, 2005; Ite, 2007) partly in response to corporate impacts on host communities and the environment (Idemudia & Osayande, 2018). This has galvanised the predominance of CSR initiatives around the development of those communities. As wider stakeholders expect greater visibility of corporate social and environmental actions through more disclosures (Eweje, 2007), the MNCs engage in broad disclosures of their CSR activities (Ite, 2004). However, those disclosures largely pass for CSR rhetoric detached from substantive CSR practices (Frynas, 2005) and are harnessed to portray the entities as transparent (Egbon et al., 2018) and responsible despite public outcries (Nwagbara & Belal, 2019). But these CSR practices in relation to community development in the Niger Delta are still emerging. For example, the CSR initiatives have evolved over time from one model to another, with each successive model seemingly attempting to overcome the deficiencies of the previous (Egbon et al., 2018; Idemudia & Osayande, 2018; Ite, 2007).

Ite (2007) discussed three CSR initiatives practised by Shell in the Niger Delta which primarily aimed at poverty alleviation, namely, community assistance (CA), community development (CD), and sustainable community development (SCD).[2] The CA operates a top-down CSR strategy with emphasis on corporate philanthropy lacking the inclusive participation of the beneficiaries, and, ultimately, resulting in ad hoc projects that are isolated from the sustainability of those communities. Unlike the CA, the CD is construed as a bottom-up CSR strategy in departure from CA's passive community participation and the execution of ad hoc projects in that it

[2] The CSR strategies/practices (including global memorandum of understanding, discussed later) adopted by Shell and other oil MNCs in Nigeria are similar but differ in nomenclature (see Chevron Nigeria, 2012; Idemudia, 2014b; Mamudu et al., 2021). However, we choose to refer to SCD as Sustainability-driven Community Development because the current usage does not appear to embed the broad notion of sustainability.

allows the benefiting communities to develop their own development plans that embed co-ordinated projects in the community interests. However, the strategy is rooted in the myth that the communities depend on the oil corporations for their socio-economic needs. Presumably on this premise, the strategy seeks to empower the communities to reduce such dependence (Ite, 2007). Unfortunately, while both the CA and CD CSR practices primarily seek to promote poverty alleviation in host communities (Ite, 2007), they pay limited attention to the negative environmental impacts associated with oil operations in the Niger Delta (Idemudia & Osayande, 2018), which largely underpin the threat to livelihood in the region (Onyena & Sam, 2020; Pegg & Zabbey, 2013).

Like the CA and CD, the SCD strategy claims to empower the benefiting communities to ably plan, design, and drive their own community development in order to improve their socio-economic life (Ite, 2007). It extends beyond the CD as it includes a mechanism for interfacing between the communities and the corporation, with the intent of fostering partnership (Ite, 2007). It signals a partnership involving the corporations, civil societies, communities, and government with a view to achieving sustainable development. An implementation of a CSR strategy that emphasised greater cooperation, transparency, and accountability underpinned the emergence of the global memorandum of understanding (GMOU) model of CSR initiative currently employed by oil corporations in Nigeria. According to Chevron Nigeria (2012, p. 16), the 'GMOU is based on the principles of Participatory Partnership and Stakeholder Engagement, Transparency and Accountability, Sustainability Assurance, Peace Building, Monitoring and Evaluation, and most importantly, community-driven development planning'. The GMOU model is an enhanced CSR initiative crafted to improve the livelihoods of the benefiting communities (Chevron, 2012; Egbon et al., 2018). But its transparency and underlying accountability in the interest of the communities are contestable due to the imbalance in corporate–community power that engenders corporate undue influence. For example, Egbon et al. (2018) noted that the release of GMOU funds is tied to an unpublicised stringent condition called the freedom-to-operate, which prohibits the communities from agitating against the corporation, carrying the consequence of forfeiting the hitherto approved funds if the hidden clause is allegedly violated. This clause is dictatorial as it restrains the agency of the communities to react against inappropriate corporate actions perceived by the communities as undermining their interests. It sums the GMOU not only as a good CSR initiative but also a subtle instrument deployed by corporations to stifle communities' power to hold them accountable for unacceptable corporate practices (Egbon et al., 2018).

Like the prior CSR initiatives, the voluntary GMOU is decoupled from the environmental impacts of the corporations, which are the major rallying point for community agitations (Denedo et al., 2019; Egbon et al., 2018; Egbon & Mgbame, 2020). Neglecting the environmental concern of the communities amounts to an abdication of negative injunction duties as best practices (Idemudia, 2010b; Nwoke, 2016). While the CSR practices in relation to community development in the Nigerian oil industry have inherent flaws, they have continued to change forms in part response to the dynamic and complex operating environment in which they are practised. Such changes are however commendable as they tend to encourage the participation of communities in community development initiatives, however marginal the level of participation. More recently, emphasis is shifting beyond the corporate voluntary initiatives such as the GMOU to a CSR initiative mandated by law. The Nigeria government has through the Petroleum Industry Act (PIA) 2021 mandated a new CSR strategy for the Nigerian oil industry (discussed later in section "Institutionalised CSR Legal Framework"). While the new model is yet to be empirically examined, this chapter examines its core provisions and offers some critiques in relation to positive affirmative duties and negative injunction duties (see sections "Critique of Prior CSR Practices: Positive Affirmative Duties/Negative Injunction Duties" and "Legal Approach to CSR and Its Implications for Corporate Accountability and Reporting").

Critique of Prior CSR Practices: Positive Affirmative Duties/Negative Injunction Duties

Generally, there is no doubt that oil corporations in Nigeria have implemented numerous CSR projects as a portrayal of their socially responsible behaviour. They have initiated and executed development schemes for the benefit of host communities. Yet, the prevailing argument appears to be that oil activities have caused much harm to the Niger Delta environment. In this regard, the contention is that the harmful consequences of oil extraction far outweigh the positive CSR projects of oil corporations (Bird, 2004). In looking at these harmful corollaries of oil extraction (as opposed to the social responsibility engagements of corporations in the area), an important distinction between two different CSR obligations has been offered (Ojo, 2012; Idemudia & Ite, 2006). The first is the positive obligation or 'affirmative duty' and the second is the duty to care for the environment or 'negative injunction duty'. The affirmative duties are those obligations which corporations perform in aid of host communities, including the building of roads, hospitals, schools, initiating and executing electrification projects, and the commissioning of water boreholes, etc. On the other hand, negative injunction duties involve preventing and rectifying social injuries that may arise in the process of realising business objectives—including oil spills, water and air pollution, gas flaring, acid rains, and forest fires, amongst others. Meeting these negative injunction duties are fundamental aspects of CSR, because they make up the 'moral minimum' which every corporation is expected to observe (Simon et al., 1972).

In recent years, the focus of the CSR debate in the Niger Delta appears to be tilting towards deconstructing the two complimentary duties of CSR: positive affirmative duties and negative injunction duties (the duty to care for the environment). The argument is that the two duties operate reciprocally and not mutually exclusive. By fulfilling the duty of care not to damage the environment, value is created. Execution of the positive affirmative duties further enhances this value that is created (Ogula,

2012). What is seen as CSR practices by corporations in Nigeria are principally affirmative in nature. The cases and general unrest in the region are, therefore, rooted primarily on the inability of corporations to fulfil the negative injunction duties. However, the corporations seemed to have over-concentrated on the positive affirmative obligations to the detriment of the negative injunction duties (Ojo, 2012; Idemudia, 2008).

This is the predominant reason why host communities in the region continue to suffer from the adverse and obnoxious social, economic, and environmental effects of oil activities (Duruigbo and others, 2001; Idemudia & Ite, 2006). For Idemudia and Ite (2006), the failure to fulfil the negative injunction duties lies at the heart of the crisis in the region. They contend that no amount of road construction or classroom renovation can offset the consequences of oil pollution, and no amount of electrification project or award of scholarships can balance the effect of gas flaring on those communities. Similarly, Ukpongson and Onu (2004) contend that it is necessary for the corporations to understand that CSR extends beyond investing in the creation of new sources of livelihood. It also includes the duty to protect and preserve existing sources of survival for host communities. No amount of positive affirmative duties will replace the duty of adherence to the moral minimum—care for the environment.

A major consequence of corporate failure to fulfil the negative injunction duty is oil spills. John Vidal (2010), the Environmental Editor of *The Observer*, undertook an investigative journey into the hinterlands of the Niger Delta to obtain first-hand information on the situation in the Niger Delta. In his report, he documented examples of oil spills in Otuegwe village—where the putrid smell of oil spill could be smelt before it is seen and where forest and farmland were now covered in sheen of greasy oil and drinking wells were polluted. According to Vidal (2010), a broken ExxonMobil pipe in Ibeno, Akwa Ibom State, led to the leak of over a million barrels of oil into the Delta Rivers, and a large oil slick was found floating on Lake Adibawa in Bayelsa State and another in Ogoniland. From existing records, there are more than 2000 principal spills and thousands more of smaller ones yet to be cleaned-up (Vidal, 2010). In this context, since 2014, Eni has reported 820 spills in the Niger Delta, with 26,286 barrels or 4.1 million litres lost (Amnesty International, 2018). Since 2011, Shell has reported 1010 spills, with 110,535 barrels or 17.5 million litres lost, which is about the size of seven Olympic swimming pools (Amnesty International, 2018). While these are massive figures, the reality may even be worse. This is because the figures reported by the corporations are significantly at variance with those reported by the Nigerian government, which recorded 1369 spills and 1659 spills for Shell and ENI, respectively, over the same period (Amnesty International, 2018).

Another grave consequence of the non-compliance with the negative injunction duties by corporations in Nigeria is the continuous flaring of gas. Gas flaring, the smouldering of natural gas associated with oil extraction, takes place due to a range of issues, from market and economic constraints to a lack of appropriate regulation and political will (see Hassan & Kouhy, 2013). The practice results in a range of pollutants released into the atmosphere, including carbon dioxide, methane, and black carbon (soot). The methane emissions from gas flaring contribute significantly to global warming in the short to medium term, because methane is over 80 times more powerful than carbon dioxide on a 20-year basis. In a report in 2021, the World Bank estimated that 142 billion cubic meters (bcm) of natural gas was flared in 2020 (World Bank, 2021). It also noted that gas flared in Africa (37 bcm in 2000) could produce 200 Terawatt hours (TWh) of power, equivalent to 50% of the current power consumption in the continent, and more than twice the level of power consumption in sub-Saharan Africa—excluding the Republic of South Africa. It is estimated that annual gas flaring is about 150 bcm (World Bank, 2004).

Globally, satellite data from 2020 reveals that Nigeria is one of the top seven gas flaring countries (World Bank, 2021). It is also estimated that 40% of gas produced in Nigeria is flared and that this accounts for about 12.5% of annual global flaring. However, these estimates are regarded as conservative by environmentalists, who believe that more realistic estimates should put gas flared in Nigeria at about 23 bcm annually, representing about 60% of production (Environmental Rights Action, 2008). However, the reported figures of gas flare have been decreasing in recent times (Hassan, 2020). Irrespective of the reported decrease in the amount of gas flared, although scientific results show that the actual amount of gas flared globally is thrice higher than the reported (The Guardian, 2022), any amount of gas flared is hazardous.

Gas flares are not only dangerous, but also have indeed produced millions of tons of both Methane and Carbon Monoxide in Nigeria (Watts, 2004). Gas flaring, in predominantly farming communities, causes acid rains which damage and prevent the pollination and growth of crops. This produces adverse consequences on the livelihood of the Niger Delta agrarian communities because a large amount of gas is simply burned off (Okeagu et al. 2006). While other countries are trying to reduce and abolish gas flaring, it still flourishes in Nigeria (Okeagu et al. 2006), the attempt by the Associated Gas Re-injection Act 1979 to abolish gas flaring from 1984 was open to discretionary interference capable of subverting its potency. In that regard, Eze (2019) contended that the 'Act did not actually abolish gas flaring but made it subject to ministerial permission', which risk compromising due to the systemic corruption in Nigeria. Despite this regulatory lax allowing corporations to flare gas with almost impunity, a community can seek redress against them from the angle of human rights violation. A case in point here is the celebrated case of Gbemre v. Shell where the plaintiff sued Shell that its gas flaring violated his right to dignity, which though he obtained a favourable court judgement against Shell, the outcome was never enforced due to corporate and political manoeuvres (May & Dayo, 2019).

Institutionalised CSR Legal Framework

Just like other developing countries and emerging economies, the concept of CSR and environmental responsibility and accountability is not new within the Nigerian business community (Amodu, 2020). CSR remains a concept that defines the responsibility companies have towards the society beyond its primary obligations to its shareholders, although largely voluntary (Amao, 2014). Many of the CSR activities in the Nigerian corporate space are associated with the MNCs operating in the oil and gas sector (Tuodolo, 2009). However, the inherent opacity and human rights violations in the Nigerian oil industry call to question the effectiveness of the CSR agenda. This raises doubt whether the MNCs' CSR initiatives are underpinned by a 'conscience clause' or a means of 'managing perceptions' when it suits them (Hackett, 2016). Despite the burgeoning CSR initiatives, the existential conflicts in the corporate–community relations in the Niger Delta have remained unobliterated, essentially as the CSR initiatives continue to relegate the negative injunction duties to blind spots. Such relegation indicates an emboldened CSR fraught with the abdication of moral obligations of the duty of care to not harm the communities and their environment, which a responsive regulatory framework ought to fix. Perhaps, this challenge led to the idea of establishing a clear CSR legal framework under the PIA 2021 for achieving Sustainability-driven Community Development (SCD) in the oil-producing communities in Nigeria.

It is important to note that a key factor that would make any established legal framework to flourish, especially in terms of improving development, is the provision of an enabling environment. In reality, this requires the leveraging policy approaches in an environment that supports and offers incentives for corporate activities that minimise environmental and social costs, while maximising economic advantages. To achieve this, the State may need to utilise an institutionalised legal framework in the specific sector, for example, the Petroleum Industry Act 2021 through the Host Community Development Trust. The purpose is to ensure that there is a system in place that places priority on SCD in the public interest of host communities as part of the CSR from the oil industry.

CSR initiatives have metamorphosed over time to the GMOU model for community development in the oil corporations in Nigeria prior to the enactment of PIA 2021 as earlier discussed. The GMOU model seeks to promote corporate–community accountability necessary to foster the SCD in Nigeria's Niger Delta (Egbon et al., 2018). In practice, this was said to be negotiated between individual operators and the respective host communities.

It is important to mention that the application of law in regulating CSR initiatives in the oil industry preceded the PIA 2021. Specifically, the Niger Delta Development Commission Act 2000 (NDDC, 2000) preceded the PIA as an indirect CSR framework to meet the developmental needs of communities in the Niger Delta where oil operations are undertaken (see Enabulele, 2020). This government-led initiative was established and saddled with the responsibility to create a master plan for the economic and infrastructural development of the oil-rich region (Agbaitoro, 2018; Enabulele, 2020).

In practice, this requires the oil corporations operating in the Niger Delta region to contribute an annual NDDC levy, which is equal to 3% of their annual budget (NDDC Act s. 14). Although the initiative is government-led, the Commission is funded by the oil corporations. As such, these corporations recognise their contribution to NDDC as part of the CSR spending (Shell, 2017). In contrast to the CSR model implemented by the state under the NDDC Act, the PIA provided for the establishment of the Host Community Development Trust (HCDT) with an operational model like the GMOU.

Chapter III of the PIA covers broad issues on Host Communities. It covers the objectives envisioned in implementing this segment of the PIA, techniques and processes required for the execution of the PIA's purposes. It also provides for how to deliver the advantages envisaged to the affected communities hosting the petroleum assets and activities in the Niger Delta. Importantly, a key part of the section is the establishment of trust funds, financing sources, and any other features that may be required. This is heavily supported by the command-and-control approach that underpins the PIA. In this regard, S. 251 (4) provides for the establishment of a trust that is based on the findings of the host community needs assessment. This is geared towards promoting SCD, specifically, taking the form of establishing a host community development plan that is to be submitted to the Nigerian Upstream Petroleum Regulatory Commission (Commission), as the case may be, in order for the Commission to undertake its oversight function prior to the establishment of the trust. In s. 251(2), the Act provides for the needs assessment of each host community from social, environmental, and economic perspectives. It is unclear though what the environmental perspective covers under the needs assessment. Does it approve more money to the communities based on the level of environmental impacts? The Act specifically contemplates the HCDT as a mechanism for community development and implicitly does not contemplate it to address corporate environmental impacts. This suggests that independent of the HCDT, the communities will have to demand the corporations to address those corporate negative environmental impacts. But the communities may unlikely be able to get corporate attention until after confrontations. This raises a caveat that puts the communities' access to the allocated HCDT funds at risk (see s. 257).

Presently, with the enactment of the PIA, under Chapter III, Section 240 (2) sets out a framework that makes CSR contribution a mandatory levy on corporations. Clearly, this approach has changed the voluntary and self-regulatory element of CSR in the industry to hard law legislation. It is expected that all voluntary host community development initiatives will now be managed under the HCDT. Under the PIA settlors, that is, a holder of an interest in a petroleum prospecting licence or petroleum mining lease or a holder of an interest in a licence for midstream petroleum operations, whose area of operations is located in or appurtenant to any community or communities, are required to set up a HCDT for the benefit of their respective host communities under s. 235 of the PIA and contribute 3% of the annual operating expenses of their previous financial year to the Trust's fund (s. 240). Failure for an oil lease/license holder to establish a HCDT within the stipulated time of 12 months and subsequent reminders might have its lease/license revoked (s. 238). Much like the freedom-to-operate clause implicit in the GMOU model, s. 257

prescribed explicit deductions against the funds accruing to the communities. So, s. 257 (2) categorically states that:

> 'Where in any year, and act of vandalism, sabotage or other civil unrest occurs that causes damage to petroleum and designated facilities or disrupts production activities within the host communities, the community shall forfeit its entitlement to the extent of the costs of repairs of the damage that resulted from the activity with respect to the provisions of this Act within that financial year: Provided the interruption is not caused by technical or natural cause'.

Whereas the above provision might appear fair to prevent sabotage or the like, it might become a subtle weapon the corporations might deploy to subjugate the poor communities given the contested nature of the causes of pollution (majorly oil spills) in the Niger Delta. For example, Egbon and Mgbame (2020) and Amnesty International (2013) have shown that the oil corporations tend to manipulatively attribute the cause of most oil spills in the Niger Delta to sabotage so they can escape liability.

The foregoing is very concerning as the conflicts between the corporations and communities have predominately come from agitations over corporate negative environmental impacts. Hence, the idea of having a clear institutionalised legal framework for CSR activities would be thought to be driven by the need for the corporations to address the challenges and grievances of the host communities. Interestingly, s. 239(3f) stated that one of the objectives of the HCDT shall include, 'to support local initiatives within the host communities which seek to enhance the protection of the environment'. Only this section and s. 251(2) are the relevant sections that linked the HCDT to the environment. S.239(3f) does not address negative injunction duties rather it can be construed as the efforts of the communities to improve their environment and not the addressing of corporate negative environmental impacts. However, it does not foreclose on the environmental or other rights of the communities. In that regard, s. 239(4) categorically established that: 'Notwithstanding the provisions of this Act relating to funding of the trust fund, nothing shall preclude the host communities from their entitlements under any other law.'

In sum, it is important to note that the successful adoption and implementation of the HCDT is meant to guarantee compulsory community-led CSR activities across the Niger Delta and other oil-producing areas in Nigeria. However, since the PIA is new, along with institutionalised HCDT CSR legal framework, it is quite early to judge the extent to which the framework will promote substantive SCD.

Legal Approach to CSR and Its Implications for Corporate Accountability and Reporting

The importance of CSR in the management and decision-making processes of corporations cannot be over-emphasised. (Aguinis & Glavas, 2012; Jenkins, 2005). As a result, beginning from decades ago, there have been numerous academic research devoted to analysing the concept of CSR, from both developed and the developing countries perspectives. (Bowen & Johnson, 1953; Davis, 1960; Friedman, 1970; Freeman, 1984, 2010; Drucker, 1984; Carroll, 2016; Panda et al., 2019). The predominant rationalisation of contemporary CSR practices is that they are based on the instrumental metaphor and self-interest (Brown & Fraser, 2006; Sorour et al., 2021). In this instrumentalist metaphor, the corporations are motivated to publicising their CSR practices to improve their public image, even when what they do is at variance with the dictates of effective CSR (Cho et al., 2015, 2018; La Cour & Kromann, 2011; Vos, 2009). The instrumentalist metaphor enables them to avoid engaging with meaningful CSR when they believe it will not contribute to the bottom line, profit maximisation (Barnett, 2016).

Crucially, the contemporary CSR is voluntary, with no enforcement mechanism where there is a breach (McInerney, 2007). Accordingly, the dearth of transparency regarding the social obligations of corporations has led to voluntary initiatives to meet obligatory responsibilities. As Tamvada (2020, p. 4) has observed, 'this voluntarism leads to core required obligations being considered as mere instruments for serving businesses resulting in misleading perception of responsibility while raising questions on the effectiveness of CSR practices'. Self-regulation has equally resulted in corporations publicising their CSR practices for strategic interests while blatantly violating human rights. For instance, Tamvada (2020, p. 4) reports that 'Volkswagen has a long list of reported CSR practices but the recent scandal over diesel emissions reveals how corporates disguise themselves as good businesses under voluntarism'. The Rana Plaza disaster in Bangladesh validates limited effectiveness of voluntarism and leads to the conclusion that there is an urgent need to critically evaluate the intrinsic flaws in contemporary CSR through a systematic process. The voluntary status accorded to CSR has impeded companies from taking proactive measures towards business responsibility engagements and presents considerable challenges towards the realisation of effective/meaningful CSR practices. Voluntarism has resulted in a stunted and sometimes opaque understandings of the limits of these engagements. This has led Osuji (2011) to argue that the dearth of mechanisms for the regulation of CSR, coupled with the tying of social issues to financial performance, has led to the stultification of independent development of the concept. Following from these, the predominant view appears to be that legal approaches to CSR is a necessity.

In looking at the legal approaches to CSR and their implication for corporate accountability and reporting, it is necessary to note that CSR has its roots in morality, which underlines the need for corporate actors, not only to comply with the positive affirmative duties, but also to fulfil the negative injunction duties (responsibility to not harm society and environment while positively contributing to the welfare of society and its stakeholders (Nwoke, 2016)). In this context, it is imperative for corporations to avoid disregarding their social responsibilities while pursuing their economic goals (Baden, 2016; Sachs et al., 2009). It is when corporations fulfil these two complimentary duties that they can be said to be socially responsible. The reality (as was shown in the preceding section) is that when it comes to the negative duties, corporations in Nigeria continue to act irresponsibly towards host communities and their environment. This has led to the clamour for these business entities to be held responsible/accountable for their actions.

Responsibility is an obligation and a duty to perform that which is supposed to be performed. Such an obligation may be ineffectively executed where there is no accountability for a failure to perform. Apparently, responsibility is essential when viewed from the context of law and accountability. According to Barry and Shaw (1979), responsibility is an aspect of duty which a person acquires because of the person's position, function, or work. From this perspective, responsibility incorporates both obligations associated with a task and the basic functions of that task. Thus, as part of responsibility, moral obligations may be related to functional obligations of a role (Bivins, 2006).

Accountability, on its own, is 'a moral or institutional relation in which entitlements are accorded to one agent (or group of agents) to question, direct, sanction or constrain the exercise of power by another' (Tamvada, 2020, p. 7; see also Macdonald, 2014). For Frink and Klimoski (1998), it is the apparent desire to validate or protect a decision or act to a group of people. The action, in this context, possesses the capacity to compensate or punish, where the compensation or punishment is seen as dependent 'on accountability conditions'. Where there is no accountability, it will be difficult to interrogate irresponsible actions of corporate actors. The consequence is that these actors become emboldened to act only to satisfy their interest. Accordingly, it can be inferred that accountability is an effective mechanism in regulating the activities of corporate actors who have the responsibility or obligations to discharge their functions in each system (Tamvada, 2020). To be accountable, an actor must be legally responsible. Put differently, legal responsibilities come with accountability (Tamvada, 2020). As Bivins (2006) argues, when harm results from an action, and there is no legitimate excuse for the action, then the party responsible should be ethically answerable for the consequences of that action. For these reasons, legal and moral responsibilities are closely connected to accountability in the context of a meaningful discharge of functions.

Furthermore, c. 255 of the PIA requires the management committee of the HCDT to submit mid-year report and annual report (including audited account) to the HCDT's Board of Trustee which would then submit the audited annual report to the settlor (the funding oil corporation). Sub-section (d) of c. 255 then requires the settlor to submit both the annual report and audit account of the HCDT to the relevant regulatory authority. It is still unclear whether these reports will be publicly available or how much of the reports the corporations will disclose in their CSR/Sustainability reports.

As already discussed, the CSR practices aimed at sustainable community development in the Niger Delta abdicate negative injunction duties which are paramount in that volatile region. The recent CSR model incorporated in the provisions of the Petroleum Industry Act (PIA) 2021 is largely undifferentiated from current CSR practices. Its HCDT operates apparently like the GMOU, having also legalised the subtle freedom-to-operate clause that constrains communities' agency to resist the violation of their rights. While the model is not aptly captured by the *'old wine in new bottle'* metaphor, it does correspond to a *'mixed wine in a new bottle'*, being a simulacrum of GMOU with legal ratification.

Concluding Remarks

This paper has evaluated the current legal regulation of corporate social responsibility practices in the Nigerian oil industry and what this means for corporate reporting. Accordingly, it has argued that CSR initiatives that conform to best practices should as a minimum prioritise the negative injunction duties as a fundamental of justice. Without a doubt, the voluntary nature of earlier CSR practices has failed to comply with these negative injunction duties. Having examined the recently passed PIA 2021, and its provisions on corporate–community relations viz-a-viz corporate responsible actions, the chapter finds that very little has changed in terms of the law holding the corporations accountable for their actions in Nigeria. In this context, the intervention submits that the legislation fails to mandate corporate compliance with the negative injunction duties, although s. 102–107 offered some provisions around environmental management and gas flaring. While provisions exist in the Act encouraging responsible practices, these, arguably, promote the positive affirmative duties. Any CSR obligation, whether voluntary or mandatory, that privileges positive affirmative duties at the expense of negative injunction duties is inconsistent with community development rooted in sustainability.

If social reporting is to be meaningful and move beyond rhetoric, then the law must insist on compliance with both the positive and the negative injunction duties. It is, therefore, imperative for the government and policy makers in Nigeria to re-evaluate the recent PIA 2021, with a view to introducing provisions in the law that will mandate compliance with the negative injunction duties in relation to promoting sustainable communities. When this is done, corporations will be mandated to report, not only the positive duties they have embarked on, but also on those harmful activities associated with their activities and how the corporations have been able to avoid them, or at least minimise their adverse impacts on host communities and their environment. It is when this is achieved that the benefits of corporate reporting anchored on legal regulation will be realised.

References

Aaron, K. K. (2010). Two theses on corporate social responsibility performance in the Niger Delta. In K. K. Aaron & G. Dawari (Eds.), *Placebo as medicine: The poverty of development intervention and conflict resolution strategies in the Niger Delta Region of Nigeria*. Kemuela Books.

Agbaitoro, G. (2018). Legal strategy for resolving the socio-economic and environmental symptoms of the resource curse in Nigeria: The role of impact and benefit agreements (IBAs). *Commonwealth Law Bulletin, 44*(3), 381–399.

Aguinis, H., & Glavas, A. (2012). What we know and don't know about corporate social responsibility: A review and research agenda. *Journal of Management, 38*(4), 932–968.

Amao, O. (2014). Emergent state practice on the creation and practice of standards on corporate social responsibility. *State Practice & International Law Journal (SPILJ), 1*(1), 117–137.

Amnesty International. (2013). *Bad information: Oil spill investigations in the Niger Delta*. Amnesty International.

Amnesty International. (2018). Decoders helped us prove how Reckless Shell and Eni are (2018). *Amnesty International Publication*. Available from https://www.amnesty.org/en/latest/news/2018/03/niger-delta-oil-spills-decoders/#:~:text=Since%202014%20Eni%20has%20reported,reality%20may%20be%20even%20worse (accessed 20 February 2023).

Amodu, N. A. (2020). *Corporate social responsibility and Law in Africa: Theories, issues and practices*. Routledge.

Baden, D. (2016). A reconstruction of Carroll's pyramid of corporate social responsibility for the 21st century. *International Journal of Corporate Social Responsibility, 1*(1), 8.

Barnett, M. L. (2016). The business case for corporate social responsibility: A critique and an indirect path forward. *Business & Society, 58*(1), 167–190.

Barry, V. & Shaw, W. H. (1979). *Moral issues in business*. Wadsworth.

Bird, F. (2004). *Wealth and poverty in the niger delta: Reflections on the history of shell's operations in Nigeria*. Concordia Press.

Bivins, T. (2006). Responsibility and accountability. In K. Fitzpatrick & C. Bronstein (Eds.), *Ethics in public relations: Responsible advocacy* (pp. 19–38). SAGE.

Bowen, H. R., & Johnson, F. E. (1953). *Social responsibility of the businessman*. Harper.

Brown, J. & Fraser, M. (2006). Approaches and perspectives in social and environmental accounting: An overview of the conceptual landscape. *Business Strategy and the Environment, 15*(2), 103–117.

Carroll, A. B. (2016). Carroll's pyramid of CSR: Taking another look. *International Journal of Corporate Social Responsibility, 1*(1), 3.

Chevron Nigeria. (2012). *Corporate responsibility report*. Lagos, Nigeria.

Cho, C. H., Laine, M., Roberts, R. W., & Rodrigue, M. (2015). Organized hypocrisy, organizational facades and sustainability reporting. *Accounting, Organizations & Society, 40*, 78–94.

Cho, C. H., Laine, M., Roberts, R. W., & Rodrigue, M. (2018). The frontstage and backstage of corporate sustainability reporting: Evidence from the Arctic National Wildlife Refuge Bill. *Journal of Business Ethics, 152*(3), 865–886.

Davis, K. (1960). Can business afford to ignore social responsibilities? *California Management Review, 2*(3), 70–76.

Denedo, M., Thomson, I., & Yonekura, A. (2019). Ecological damage, human rights and oil: Local advocacy NGOs dialogic action and alternative accounting practices. *Accounting Forum, 43*(1), 85–112.

Drucker, P. F. (1984). Converting social problems into business opportunities: The new meaning of corporate social responsibility. *California Management Review, 26*(000002), 53.

Duruigbo, E., Wozniak, O., & Leighton, M. (2001). *Oil development in Nigeria: A critical investigation of chevron corporation's performance in the niger river delta*. Berkeley.

Egbon, O., Idemudia, U., & Amaeshi, K. (2018). Shell Nigeria's Global Memorandum of Understanding and corporate-community accountability relations: A critical appraisal. *Accounting, Auditing & Accountability Journal, 31*(1), 51–74.

Egbon, O., & Mgbame, O. C. (2020). Examining the accounts of oil spills crises in Nigeria through sensegiving and defensive behaviours. *Accounting, Auditing & Accountability Journal, 33*(8), 2053–2076.

Enabulele, O. (2020). Mitigating the effects of environmental degradation in the oil industry: An assessment of government compensation scheme in Nigeria. *Law, Environment and Development Journal (LEAD Journal), 16*(1), 76–92.

Environmental Rights Action. (2008). *Fact Sheet: Harmful Gas Flaring in Nigeria* (2008). <http://www.foe.org/pdf/GasFlaringNigeria_FS.pdf> (accessed 23 February 2023).

Eweje, G. (2007). Multinational oil companies' CSR initiatives in Nigeria The scepticism of stakeholders in host communities. *Managerial Law, 49*(5/6), 218–235.

Eze, T. (2019). Updating the legal framework for the elimination of gas flaring in Nigeria. *African Journal of Criminal Law and Jurisprudence, 4*, 183–192.

Freeman, R. E. (1984). *Strategic management: A stakeholder approach*. Pitman.

Freeman, R. E. (2010). *Strategic management: A stakeholder approach*. Cambridge University Press.

Friedman, M. (1970). *The social responsibility of business is to increase its profits*. New York Times Magazine.

Frink, D. D., & Klimoski, R. J. (1998). Towards a Theory of Accountability in organisations and human resource management. In G. R. Ferris (Ed.), *Research in personnel and human resource management* (Vol. 16, pp. 1–51). JAI Press.

Frynas, J. G. (2005). The false developmental promise of corporate social responsibility: Evidence from multinational oil companies. *International Affairs, 81*(3), 581–598.

Frynas, J. G., & Stephens, S. (2015). Political corporate social responsibility: Reviewing theories and setting new agenda. *International Journal of Management Reviews, 17*(4), 483–509.

Gray, R. (2006). Does sustainability reporting improve corporate behaviour?: Wrong question? Right time? *Accounting and Business Research, 36*(S1), 65–88.

Hackett, C. (2016). The challenge of MNCs and development: Oil extraction, CSR, Nigeria and corruption. *Journal of Human Rights in the Commonwealth, 2*(2), 8.

Hassan, A. (2020). Dealing with vulnerability to carbon emission from gas flaring: The roles of transparency and utilisation policies in Nigeria. *OPEC Energy Review, 44*(4), 369–403.

Hassan, A., & Kouhy, R. (2013). Gas flaring in Nigeria: Analysis of changes in its consequent carbon emission and reporting. *Accounting Forum, 37*(2), 124–134.

Hussain, W., & Moriarty, J. (2018). Accountable to whom? Rethinking the role of corporations in political CSR. *Journal of Business Ethics, 149*(3), 519–534.

Idemudia, U. (2008). Conceptualising the CSR development debate: Bridging existing analytical gaps. *Journal of Corporate Citizenship, 28*, 101.

Idemudia, U. (2009). Oil extraction and poverty reduction in the Niger Delta: A critical examination of partnership initiatives. *Journal of Business Ethics, 90*((1) S1), 91–116.

Idemudia, U. (2010a). Rethinking the role of corporate social responsibility in the Nigerian oil conflict: The limits of CSR. *Journal of International Development, 22*(7), 833–845.

Idemudia, U. (2010b). Conceptualising the CSR and development debate: Bridging existing analytical gaps. In P. Maiti (Ed.), *Corporate social responsibility: Critique, policies and strategies* (Vol. 1, pp. 65–96). Sharda Publishing House.

Idemudia, U. (2014a). Oil companies and sustainable community development in the Niger Delta, Nigeria: The issue of reciprocal responsibility and its implications for corporate citizenship theory and practice. *Sustainable Development, 22*(3), 177–187.

Idemudia, U. (2014b). Corporate-community engagement strategies in the Niger Delta: Some critical reflections. *The Extractive Industries and Society, 1*, 154–162.

Idemudia, U., & Ite, U. E. (2006). Corporate-community relations in Nigeria's oil industry: Challenges and imperatives. *Corporate Social Responsibility and Environmental Management, 13*(4), 194–206.

Idemudia, U., & Ite, U. E. (2010). Corporate-community relations in Nigeria's oil industry: Challenges and imperatives. In P. Maiti (Ed.), *Corporate social responsibility: Critique, policies and strategies* (Vol. 1, pp. 44–64). Sharda Publishing House.

Idemudia, U., & Osayande, N. (2018). Assessing the effect of corporate social responsibility on community development in the Niger Delta: A corporate perspective. *Community Development Journal, 53*(1), 155–172.

Ite, U. E. (2004). Multinationals and corporate social responsibility in developing countries: A case study of Nigeria. *Corporate Social Responsibility and Environmental Management, 11*(1), 1–11.

Ite, U. E. (2007). Changing times and strategies: Shell's contribution to sustainable community development in the Niger Delta, Nigeria. *Sustainable Development, 15*(1), 1–14.

Jenkins, R. (2005). Globalization, corporate social responsibility and poverty. *International Affairs, 81*(3), 525–540.

La Cour, A., & Kromann, J. (2011). Euphemisms and hypocrisy in corporate philanthropy. *Business Ethics: A European Review, 20*(3), 267–279.

Macdonald, K. (2014). The meaning and purposes of transnational accountability. *Australian Journal of Public Administration, 73*(4), 426–436.

Mamudu, A., Mamudu, A., Elehinafe, F., & Akinneye, D. (2021). Recent trends in corporate social responsibilities in Nigeria – A case study of major oil firms in the Niger delta region. *Scientific Africa, 13*, e00928.

Matten, D., & Moon, J. (2020). Reflections on the 2018 decade award: The meaning and dynamics of corporate social responsibility. *Academy of Management Review, 45*(1), 7–28.

May, J. R., & Dayo, T. (2019). Dignity and environmental justice in Nigeria: The case of Gbemre v. Shell. *Widener Law Review, 25*(2), 269–284.

McInerney, T. (2007). Putting regulation before responsibility: Towards binding norms of corporate social responsibility. *Cornell International Law Journal, 40*, 171.

Muthuri, J. N., Moon, J., & Idemudia, U. (2012). Corporate innovation and sustainable community development in developing countries. *Business & Society, 51*(3), 355–381.

NDDC (2000). Niger Delta Development Commission (Establishment etc) Act. 2000, No. 6, Law of the Federal Republic of Nigeria.

Noah, A. O., Adhikari, P., Ogundele, B. O., & Yazdifar, H. (2021). Corporate environmental accountability in Nigeria: An example of regulatory failure and regulatory capture. *Journal of Accounting in Emerging Economies, 11*(1), 70–93.

Nwagbara, U., & Belal, A. (2019). Persuasive language of responsible organisation? A critical discourse analysis of corporate social responsibility (CSR) reports of Nigerian oil companies. *Accounting, Auditing and Accountability Journal, 32*(8), 2395–2420.

Nwoke, U. (2016). Two complimentary duties under corporate social responsibility Multinationals and the moral minimum in Nigeria's Delta region. *International Journal of Law and Management, 58*(1), 2–25.

Ogula, D. (2012). Corporate social responsibility: Case study of community expectations and the administrative systems, Niger Delta. *The Qualitative Report, 17*(73), 1–27.

Ojo, G. U. (2012). Community perception and oil companies' corporate social responsibility initiative in the Niger Delta. *Studies in Sociology of Science, 3*(4), 13–14.

Okeagu, J. E., Okeagu, J. C., Adegoke, A., & Onuoha, C. (2006). The environment and social impact of petroleum and natural gas exploration in Nigeria. *Journal of Third World Studies, 13*(1), 199–218.

Onyena, A. P., & Sam, K. (2020). A review of the threat of oil exploitation to mangrove ecosystem: Insights from Niger Delta, Nigeria. *Global Ecology and Conservation, 22*, e00961.

Osuji, O. (2011). Fluidity of regulation-CSR nexus: The multinational corporate corruption example. *Journal of Business Ethics, 103*(1), 31–57.

Panda, S., D'Souza, D. E., & Blankson, C. (2019). Corporate social responsibility in emerging economies: Investigating firm behavior in the Indian context. *Thunderbird International Business Review, 61*(2), 267–276.

Pegg, S., & Zabbey, N. (2013). Oil and water: The Bodo spills and the destruction of traditional livelihood structures in the Niger Delta. *Community Development Journal, 48*(3), 391–405.

Sachs, S., Rühli, E., & Kern, I. (2009). *Sustainable success with stakeholders: The untapped potential*. Springer.
Scherer, A., Palazzo, G., & Matten, D. (2009). Introduction to the special issue: Globalization as a challenge for business responsibilities. *Business Ethics Quarterly, 19*(3), 327–347.
Scherer, A. G., & Palazzo, G. (2011). The new political role of business in a globalized world: A review of a new perspective on CSR and its implications for the firm, governance, and democracy. *Journal of Management Studies, 48*(4), 899–931.
Shell. (2017). *Shell sustainability report 2016*. Royal/Dutch Shell, Netherlands.
Simon, G. J., Powers, W. C., & Gunnemann, P. J. (1972). The responsibilities of corporations and their owners. In T. L. Beauchamp & N. E. Bowie (Eds.), *Ethical theory and business* (pp. 60–75). Prentice-Hall.
Sorour, M. K., Shrives, P. J., El-Sakhawy, A. A., & Soobaroyen, T. (2021). Exploring the evolving motives underlying corporate social responsibility (CSR) disclosures in developing countries: The case of "political CSR" reporting. *Accounting, Auditing & Accountability Journal, 34*(5), 1051–1079.
Tamvada, M. (2020). Corporate social responsibility and accountability: A new theoretical foundation for regulating CSR. *International Journal of Corporate Social Responsibility, 5*(2), 1–14.
The Guardian. (2022). *Oil and gas greenhouse emissions 'three times higher' than producers claim* (Wednesday 9 November 2022). The Guardian. Available from https://www.theguardian.com/environment/2022/nov/09/oil-and-gas-greenhouse-emissions-three-times-higher-than-producers-claim (accessed 3 March 2023).
Tuodolo, F. (2009). Corporate social responsibility: Between Civil Society and Oil and Industry in the Developing World 8ACME. *International E-Journal for Critical Geographies* 530–541, 531.
Ukpongson, M., & Onu, D. (2004). Development efforts of oil companies as perceived by rural households in selected oil producing communities of Rivers State, Nigeria. *Journal of Agriculture and Social Research, 4*(1), 47–61.
Vidal, J. (2010). Nigeria's Agony Dwarfs the Gulf oil spill-the US and Europe ignore it (Sunday 30 May 2010). *The Observer Newspaper*. Available from www.guardian.co.uk/world/2010/may/30/oil-spills-nigeria-niger-deltashell (accessed 26 February 2023).
Vos, J. (2009). Actions speak louder than words: Greenwashing in corporate America. *Notre Dame Journal of Law Ethics & Public Policy, 23*, 673.
Watts, M. (2004). Resource curses? Governmentality, oil and power in the Niger Delta. *Geopolitics, 9*(1), 50–80.
World Bank. (2004). *Global gas flaring reduction initiative: Report on consultations with stakeholders, the World Bank, and the Government of Norway*. Available from https://documents1.worldbank.org/curated/en/297051468762607776/pdf/272750global1g1tiative0Report0no-01.pdf (accessed 27 February 2023).
World Bank. (2021). *Seven countries account for two-thirds of global gas flaring* (2021). World Bank Publication. Available from https://www.worldbank.org/en/news/press-release/2021/04/28/seven-countries-account-for-two-thirds-of-global-gas-flaring (accessed 20 February 2023).

Osamuyimen Egbon, PhD St Andrews, is a Lecturer in Accounting at the Essex Business School, University of Essex, UK. His research emerges from the intersection of organizations, sustainability and society and he explores the topics of accountability, social and environmental reporting, NGO accountability and community engagement, corporate social responsibility, and sustainability (including the Sustainable Development Goals). His recent works have focused on the accountability relations of multinational corporations and host communities in developing economy context. He is also currently interested in accounting for water in relation to meeting sustainable development goals, especially in developing countries and how water-intensive corporations account for water in developing countries. He is interested in accounting for disasters/crisis and human rights, as well as accounting for prisons.

Uchechukwu Nwoke, PhD, is a Lecturer at Essex Law School. Dr Nwoke's research focuses on the relationship between law and business, and how economic factors impact business operations. Specifically, he is interested in exploring how law and business interact, and how a fair and structured approach supported by rules is essential to ensure that companies operate efficiently, ultimately contributing to the economic growth of society. Before joining the University of Essex, Dr Nwoke was a Senior Lecturer at the Faculty of Law, University of Nigeria, Enugu Campus. He holds a PhD from the University of Kent, Canterbury, UK, an LLM from the University of Wales, UK, and an LLB from the University of Nigeria. He was called to the Nigerian Bar in 2005. Dr Nwoke has also worked as a consultant for NGOs in Nigeria, working with the UNDP to train parties on treaty bodies and obligations of state parties under treaties, and monitoring follow-up obligations on human rights monitoring and reporting mechanisms.

Godswill Agbaitoro, PhD, is a lecturer at Essex Law School and his research primarily relates to implementing the energy justice framework in Africa and the Global South. His PhD research funded by the University of Essex explores the extent to which the principles embedded in the energy justice framework can be deployed as a strategic tool for addressing energy access and poverty challenges, including the articulation of providing just energy systems. Godswill was a Visiting Research Scholar in 2019 at the University of Houston Law Center (UHLC), Texas, USA. His prior work experience includes practising as a lawyer following his qualification as a Barrister and Solicitor in Nigeria in 2011, working as a research assistant to the Attorney-General & Commissioner for Justice, Edo State, Nigeria, and serving as a peer reviewer for the Journal of African Law (Cambridge University Press), and Commonwealth Law Bulletin (Routledge). He received a Master of Laws (LLM), Oil and Gas Law, 2014, University of Aberdeen, UK, and a Bachelor of Laws (LLB), 2010, Madonna University, Nigeria. He is presently a member of the Nigerian Bar Association, the Society of Legal Scholars, an Associate Editor of the Journal of Sustainable Development Law and Policy, and a Fellow of the Higher Education Academy. Dr Godswill's teaching and research interests include international environmental law and sustainability, energy justice and transition, corporate social responsibility, climate justice, natural resource governance, contract law, and foundations of public law.

Chapter 2
Humanism as an Institutional Driver of Corporate Social Responsibility (CSR) Activities

David Katamba, Christopher M. J. Wickert, and Cedric M. Nkiko

Background

The current Corporate Social Responsibility (CSR) debates (Katamba & Wickert, 2021; Idowu, 2021) have highlighted a need for reforms in corporate organizations' behaviour and intentions. This strikes a counterbalance to earlier studies which centred on profit maximization (Friedman, 1970), competitiveness (Porter & Kramer, 2002, 2006), risk migration, compliance, and opportunity maximization (Grayson & Hodges, 2004), reputation (Fombrun & Van-Reil, 2004), and corporate philanthropy (Baskin, 2006; Asongu, 2007).

With each side of CSR (*Supportive and Opposing side*) taking a posture, institutionalizing CSR needs to be given priority in this era of sustainable development, but in a sense which recognizes and traces how CSR entered/was introduced in a particular organization. This is because despite CSR's importance (ISO 26000, Caroll, 1979, 1999; Bowen, 1953; Davis, 1967), in the context of sustainable development, we need to investigate the personality of the individual "human being" (humanism) behind CSR activities, which need was once mentioned by scholars of CSR (Hemingway & Maclagan, 2004; Melé & Castella 2010). Particularly, in the realm of institutionalizing CSR, these scholars left unanswered questions like, *"can a corporation be regarded as a moral agent, or should [its] moral actions be*

D. Katamba (✉)
Makerere University Business School (MUBS), Kampala, Uganda
e-mail: dkatamba@uccsri.com; dkatamba@mubs.ac.ug

C. M. J. Wickert
Vrije Universiteit (VU) Amsterdam, Amsterdam, Netherlands
e-mail: christopher.wickert@vu.nl

C. M. Nkiko
Worcester University, London, UK

U. Nwagbara et al. (eds.), *Corporate Social Responsibility Disclosure in Developing and Emerging Economies*, CSR, Sustainability, Ethics & Governance, https://doi.org/10.1007/978-3-031-61976-2_2

acknowledged as behaviour and responsibility of an individual human being"? (Hemingway & Maclagan, 2004, p. 40). Answering this and other related questions raised by Bowen (1953) when he torched through the responsibilities of a businessman, is the purpose of this publication. Additionally, as efforts keep mounting in the field of "Institutionalizing, Governing and Regulating CSR", investigating the "human nature/human being" will reveal whether CSR activities can be attributed to human beings (individuals), should be better understood as corporate behaviour, or is a combination of both. Resultantly, providing such findings (in this publication) will propel and guide scholarship and further debate into the *"humanism"* of CSR activities.

To deliver the purpose of this publication (study), we are combining knowledge on extant CSR evolution and management literature. Additionally, since we are unearthing human nature and behaviours behind CSR, we also borrowed knowledge from psychology studies (Gilbreth, 1914; Follet, 1924, 1927; Maslow, 1954; Bugental, 1964; Baier, 1993; Greening, 2006). This borrowing was necessary because psychology is the scientific study of human behaviour and mental processes which can explain why human beings behave the way they do (in this particular case, why they choose/chose to undertake particular CSR activities or why they push/pushed to institutionalize them). Henceforth, this publication was guided by one main research objective:

> "To examine how human nature [*humanism*] explains the choice, institutionalization, and sustainability of CSR activities undertaken by an organization".

In this paper, developing countries, particularly Uganda, have been preferred because they share similar characteristics which call for CSR interventions. These include chronic poverty, unprecedented corruption, human development challenges (United Nations Economic and Social Council, 2020; Ritchie & Mispy, 2018), and political development challenges (Connelly, 2010). To this end, they need almost related approaches to CSR activities (Visser, 2008b). This paper's contributions to the debate on the "humanism of CSR activities", are derived from data about the founders/owners of these companies (regardless of domicile), and the managers in charge of CSR activities.

Definition of Humanism and CSR

This research recognizes numerous conflicting CSR definitions and meanings (Bowen, 1953; Carroll, 1979, 1999; EC, 2001; World Bank Institute, 2002; Visser, 2008a; WBCSD, 2000; White, 2006; ISO 26000). Hence, we borrowed a simplified meaning (Katamba et al., 2012) which could fit our study of CSR in a sub-Saharan country. That is, CSR is "a business decision making process, ranging through business operations and discrete practices via a set of integral policies, practices, and programs that reflect a business' commitment to improving the quality of life of the workforce, the overall relationships with numerous stakeholders, and accountability to them as well as the wider community" (p. 4). Similarly, the concept of "*humanism*" is very old, though it has been under-researched, especially when it comes to integrating it in CSR debates (Melé & Castella, 2010). Its meaning and interpretation may be ascertained both from rationalist and religious (Vatican Council II, 1965; PCPJ, 2004) perspectives, although the former is more apt. The Atheist Empire website defines humanism as a "variety of ethical theory and practice that emphasizes reason, scientific inquiry, and human fulfilment in the natural world and often rejects the importance of belief in God". Similarly, Quizlet defines it as "any system or mode of thought or action in which human interests, values, and dignity predominate". Further review of literature (Melé & Mammoser, 2011; Melé & Castella, 2010; Pope John Paul II, 1991; Vatican Council II, 1965) indicates that "*humanism*" and "*humanistic*" are related terms and may be used interchangeably. However, critical observation, more so from the works of Melé and Mammoser (2011) indicates these as being different. That is, "*humanism*" sees an individual human being "as a giver" while "humanistic" sees an individual human being "as a receiver and ultimate beneficiary". In fact, Melé and Mammoser coined the term, "*Humanistic Corporate Community Involvement (HCCI)*", which in essence reflects how a firm (or organization) would retrospectively (through Corporate Community Investment) ensure integral human development of everyone in that community when it is making, say, a donation, contributing goods and services to that community, etc. However, literature about SMEs (Nkiko, 2009) reveals that humanism means that the individuals behind the organization are in essence the firm itself, since it is hard to differentiate between the firm and the people behind it. Thus, in this paper, we take the position that the individuals behind organizations/ companies are the ones who "give", though it is their organizations that implement the "giving", a position which we call "humanism".

Humanism and the Choice of CSR Activity— Historical Perspective

Psychologist Gilbreth (1914), and, management consultant, Follet (1896, 1924, 1927) contend that the practice and reflection of a human side into the betterment of workers, communities, the general business operating conditions, and the natural environment that sustains the individual interest of businessmen, is quite old. However, they are reluctant to specify how "old" it might be. In this regard, Heald (1957, 1970) says that in the 1800s, the owners of R. H. Macy Company of New York City contributed to charity and social causes. However, following several audits and court suits, in 1887, the owners were compelled to list such contributions under the miscellaneous expenses of the company's books of accounts. On a related note, Management historians (Wren, 2005; Graham, 2000) identify industrialist John H. Patterson of National Cash Register as one of the fundamentalists who provided a more humanistic reflection of factories (and businesses) as "a source of numerous social and environmental problems (like labour unrest, slums, poverty, child labour, pollution, etc.)" He therefore urged his colleagues to reflect on humanism in doing business and proposed creative ways in which to do this. In more specific terms, he suggested reflections that would contribute to solving business-generated problems (for instance, construction of hospitals/clinics, recreational facilities for employees and the surrounding communities, lunch hours/rooms, etc.). As a result, in 1893, the owner of Pullman Palace Car Company, George M. Pullman, created a model industrial community town from his personal contribution, but through his company. This town portrayed a good business policy pursued by the Palace Company owners and managers (Heald, 1970). Mr. Pullman's actions testified his humanism in improving the capacity for attracting and retaining employees, as well as improving the living standards not only for the people who were directly generating wealth for him (his employees) but also for the whole community. Other philanthropic individuals during the 1800s include Cornelius Vanderbilt (Stiles, 2009; Carey, 2005) and John Davison Rockefeller (Chernow, 1998) whose individualistic initiatives towards their employees' betterment and the general community, as well as overall business and national development, thrive to this day and still inform highly organized CSR approaches such as academic institutions like Vanderbilt University, and foundations like The Rockefeller Foundation.

Humanism and Choice of CSR Activities—Recent Debates

In his power-responsibility equation, Davis (1967, p. 48) notes that the "social responsibility of businessmen arises from the amount of social power they have". He further indicates that humans who do not take responsibility for their power shall lose it. Interestingly, he applied his thinking and successfully embedded it in the debate on CSR, saying that "society grants legitimacy and power to business and in the long run, those who do not use the granted power in a manner which the society considers responsible will lose it" (p. 314). We also see that Davis' arguments are based on human values and responsibility, social legitimacy and having a vision. However, because businessmen have different individual orientations (Katamba & Kekaramu, 2012), we amplify Davis' earlier views with Keim's (1978) "self-interest assumption and group behavior" to fit humanism in understanding the CSR undertakings of the corporations that these businessmen run. Keim (1978) indicates that in a scenario where there are different forms of CSR behaviour and undertakings, such behaviour can be explained by re-examining the implications of the individual self-interest assumption for group behaviour. He observes that several individuals behind the company have several self-interests. These, however, must be merged into a common interest that could be advanced by a collective effort (i.e. a company or institution). This common interest is what this research terms "CSR vision/philosophy" and is later operationalized and branded "corporate social responsibility".

Similarly, Visser (2008b) points out that in developing countries, CSR is largely rooted in ethical investment issues relating to, for example, apartheid and ethics of South African managers (for the case of South Africa), community involvement, socially responsible production, and employee relations (in Malaysia), child labour and ethics (in Thailand), and non-discrimination and equal opportunities (in Hong Kong). While Visser's works informed our study, the limitation was that the motives therein are gathered from largely extrinsic (e.g. employees, communities, suppliers, law enforcers, etc.), rather than intrinsic, (revealing what is inside the mind of the CSR managers or business owners) drivers. We therefore complemented Visser's works with some relevant studies by Melé and Castella (2010), Melé and Mammoser (2011), Hemingway and Maclagan (2004), and Katamba et al. (2012) that present a somewhat natural fit of humanism in the debate on CSR. For example, Melé and Mammoser (2011) largely used religious teachings to point out humanistic intentions, while Hemingway and Maclagan (2004) reflected more on human values.

Setting the Stage to Investigate "Humanism" As an Institutional Driver of CSR Activities

Using "*historical case study approach/methodology*"[1] as suggested by Widdersheim (2018),[2] we gathered data from four prominent large companies (employing more than 100 employees) which have footprints (operations) in Uganda. These are: Toyota, Unilever, Bewley's Coffee and Tea Company, and Spear Group of Companies). This small number was chosen to allow for a more detailed investigation into the "humanism of CSR activities" than would have been possible if we had used a large sample size. In addition, these companies were chosen because they have footprints not only in Uganda, but also elsewhere in the world, especially in developing countries (save for Spear Group of Companies, which is purely a Ugandan company but it is a franchise of the former Germany's Daimler Chrysler and now, The Mercedes-Benz Group AG).[3] "Footprints" mean that they are either a subsidiary of other companies, an agent, a franchise, or a supplier to other companies outside Uganda. This enabled us to compare how humanism has manifested

[1] Historical case study involves analysing cases from the distant past to the present, using eclectic data sources, to produce both idiographic and nomothetic knowledge.

[2] Michael M. Widdersheim, Assistant professor in the School of Library and Information Management at Emporia State University in Kansas, USA. He received his PhD from the School of Information Sciences at the University of Pittsburgh. He has published in *Journal of Documentation, Library and Information Science Research, Library Review, Library Trends,* and *Libri.*

[3] The Mercedes-Benz Group AG (previously named Daimler-Benz, DaimlerChrysler, and Daimler) is a German multinational automotive corporation headquartered in Stuttgart, Baden-Württemberg, Germany. It is one of the world's leading car manufacturers. Daimler-Benz was formed with the merger of Benz & Cie. and Daimler Motoren Gesellschaft in 1926.

and shaped CSR in these companies' parent companies (except for Spear Group of Companies), as well as trace some sort of pattern in their CSR activities. Lastly, we assumed that these companies are committed to CSR, given the years (over 25) for which they had practised it.

Preference was given to historical CSR data as opposed to most recent facts. This helped us to excavate and critically reflect on the personality of the founders, their successors, or managers behind the CSR activities of that time, of the sampled companies. Additionally, this preference was meant to guard against the tendency to view companies as agents *(using the institutional veil, or based on Agency Theory)*, and to reinforce the fact that the individuals behind organizations can indeed make a difference, but through their organizations. To access this data, we applied data-mining techniques, as well as requesting senior staff in these organizations to provide us with relevant records. The data sought was mainly from secondary sources (newspapers, conference/events video transcripts, organizational websites, the organizations' libraries/resource centres, authentic website downloads and postings, letters and company public documents like sustainability or annual reports).

Case study methodological approach to data collection (McLeod, 2008; Sjoberg et al., 1991) was used. Guided by CAPAM (2010), we employed a combination of: (i) the learning history approach, where we collectively reflected on the experiences of the individuals (the founders, their successors, and managers in charge of CSR) behind the organizations' CSR activities to draw constructive lessons. This enabled us to analyse actions, events, and episodes from a variety of points of view; (ii) the best practice approach, where we placed emphasis on analysing worthwhile and replicable CSR practices and human mindsets. We also borrowed from Stake's (1995, 2000) intrinsic approach and Yin's (1994) illustrative case study method to provide a descriptive account of the main characteristics of these humans advancing CSR activities.

Triangulation was used to establish data validity and reliability (Stake, 1995). For construct and internal validity of the individual cases, we used guidance from Levy (1988). Given that we studied multiple cases, we followed Yin (1984, 1994) to establish the validity and reliability of the overall findings. Meanwhile, the fact that case study data is largely qualitative (Johansson, 2003; Tellis, 1997), we took guidance from Miles and Huberman (1984), plus Yin's (1994) analytical strategy, to analyse the data. That is, we employed pattern matching, explanation building, and time series analysis to develop a framework ("THE HUMANISM OF CSR MODEL") within which the gathered data could be condensed and analysed. For generalization purposes, we were guided by the adaptive theory approach suggested by Layder (1998).

Case Studies

The four case studies evidenced below are drawn from both primary and secondary data. That is: (i) Toyota Motor Corporation (a Japanese based multinational motor car manufacturing company) and its franchise Toyota Uganda Ltd. (*now*, CFAO Motors Uganda); (ii) Unilever Plc (a worldwide manufacturer and distributor of personal care and household products in over 190 countries) and its subsidiary Unilever Uganda Ltd.; (iii) Bewley's Coffee and Tea Company (an international Ireland-based coffee and tea merchant company found in 1835 by Samuel Bewley and his son Charles. It also has operations in Uganda. Currently it is managed by Patrick Bewley, great-great grandson of Ernest Bewley); and (iv) Spear Group of Companies (a family of companies including manufacturing, importing, media, and enamelling).

(i) *Toyota Motor Corporation Japan,*[4] *and Toyota Uganda Ltd (CFAO Motors Uganda)*[5]

- *CSR activities*[6]

With a worldwide average CSR annual financial allocation of ¥ 11 billion (though approximately¥ 12.1 billion and ¥ 13.9 billion was spent in 2009 and 2010, respectively), Toyota conducts independent social contribution activities. All these CSR activities, however, fall into three focus fields –environment, traffic safety, and education, even though there is some flexibility to add other fields as country-specific societal needs may dictate. For example, in Japan, society and culture has been added. In undertaking the CSR activities, emphasis is also placed on employee volunteer activities to utilize Toyota's expertise and resources. Although Toyota established the Toyota Foundation in 1974 to act as its grant-making body to help create

[4]Toyota Motor Corporation is a Japanese multinational automotive manufacturer headquartered in Toyota City, Aichi, Japan. It was founded by Kiichiro Toyoda and incorporated on August 28, 1937. Toyota is one of the largest automobile manufacturers in the world, producing about ten million vehicles per year. https://global.toyota/en/company/

[5]At the time of collecting information for this publication, Toyota Uganda Ltd. had not yet changed its name to CFAO MOTORS UGANDA. This change took place on Wednesday, second February 2022. CFAO Motors Uganda, formerly Toyota Uganda, is a leading automotive company in Uganda. The company is a subsidiary of CFAO Automotive, a division of the CFAO Group, which is part of Toyota Tsusho Corporation. https://www.cfaogroup.com/en/toyota-uganda-limited-becomes-cfao-motors-uganda/

[6]The CSR activities and management processes at Toyota Motor Corporation are far more complex, and one may therefore be tempted to think that they cannot apply to developing countries, but they do indeed apply. They are all geared towards sustainable development. Visit the "overview of Toyota's CSR activities" on http://www.toyota-global.com/sustainability/csr_initiatives/csr_concepts/csr_activities.html.

a people-focused society and increase human happiness due to Toyota's presence in business, it also runs country-specific CSR activities.[7] That is:

(a) In environment, Toyota runs the TOYOTA Mie Miyagawa Forest Project, Toyota Environmental Activities Grant Program, research, other activities related to global warming countermeasures and biodiversity conservation, volunteer activities to preserve loggerhead turtle spawning beaches (all in Japan), and the "Eco-Schools" Environmental Program (in Europe).
(b) In traffic safety, it runs traffic safety campaigns, the Toyota Driver Communication programme, hands-on traffic safety events (all in Japan); as well as Toyota Driving Expectations, a scheme to promote traffic safety education for teens (in the USA); Supporting Road Safety Education for Young People through Toyota and You (in Argentina); children's traffic awareness activities (in Turkey); and the Toyota Safety Education Program (in India).
(c) In education, Toyota runs the Toyota Technology Challenge, which supports science and technology education (in the UK); the Toyota Study Assistance Fund, which supports Chinese Students (in China); scholarships for the children of model drivers (in Taiwan); and Basic Academic Skills–Toyota Teach (in South Africa).
(d) In society and culture, Toyota runs Net TAM, an Art Management Site (in Japan); the Pinta Tu Escuela (Paint your School), programme (in Venezuela); Toyota Production System Support Center (in the USA); Providing Regular Medical Service in Impoverished Areas (in the Philippines); Toyota Concert Tour Vietnam, which spreads the Joy of Classical Music (in Vietnam); and, lastly, Natural Disaster Relief (in Japan and overseas).

Toyota Uganda Ltd.'s (CFAO MOTORS UGANDA's) CSR centres on five themes: road safety, customer relations, health and wellness, environmental protection, and ethical trading (especially reducing counterfeits in the market).[8] CSR activities for the period 2008–2011 included partnering with Tullow Oil to promote health in the Albertin region of Uganda; support to the School of the Deaf; mosquito-net distribution in Northern Uganda to control malaria; and organizing corporate CSR training at Bonita International School. CFAO MOTORS UGANDA also mobilized corporate organizations through a golf tournament to raise money for improving the water and sanitation system for Sanyu Babies Home (a place for homeless and abandoned babies in Uganda). The money raised complemented

---

[7] Toyota's country-specific CSR activities follow the "Toyoda Precepts" established in 1935, which extol principles such as "Be contributive to the development and welfare of the country" and "Be reverent, and show gratitude" based on Sakichi's teachings. The CSR spirit of Toyota is continuously passed on to new generations. "The Toyota Global Vision", which was announced in March 2011, is a distillation of our resolve at Toyota—constantly drive "always better cars" to exceed customer expectations and contribute to communities in ways that enrich people's lives (see pp. 2 of the Toyota 2011 Sustainability Report at: http://www.toyota-global.com/sustainability/report/citizenship/pdf/citizenship.pdf).

[8] Sustainability | Toyota Motor Corporation Official Global Website, https://global.toyota/en/sustainability/?padid=ag478_from_header

CFAO MOTORS UGANDA's funds to demolish the old water tank at this home, which was harvesting dirty water, and instituted a brand new 5000 litre, automated tank.

Notably, Toyota has no uniform approach to its CSR activities. Its approach revealed huge combinational traces of: philanthropy (cash donations, motor vehicle donations, etc.,) in projects that may not directly relate to its core and peripheral business of motor manufacturing; corporate social investment (CSI) for CSR activities that directly contributes to its core business (where it sets up plants in selected places, especially for CSR activities that involve motor technology transfer and innovation); and strategic business engagement (for instance, in traffic education, environmental safety, etc.) where CSR activities relate to business functions like marketing, sales, or distribution, which are not the core business of Toyota.

- *Humanism within the above CSR activities*

According to the Toyota (2011), before 1890, Sakichi Toyoda saw the difficulties his mother encountered in operating her inefficient manual loom. This experience prompted him to devise and invent means of helping her. Later, in 1925, Sakichi furthered his desire to support inventions that enrich people's lives, to the extent that he donated ¥one million to the Imperial Institute of Invention and Innovation, which was intended to promote the development of rechargeable batteries. Sakichi devoted his life to "research and development" as his contribution to a better society (Toyota, 2011). His experiences, spirit, and orientation were inherited by Kiichiro Toyoda, the founder of Toyota Motor Corporation, and further into later generations Toyodas, to the extent that it is currently followed by Toyota and its subsidiaries (Toyota Motor (China) Investment, Toyota Motor North America, Toyota Motor Europe, and Toyota Motor Asia Pacific), who are now referred to as the "Toyoda Precepts" (Toyota, 2011). Although established in 1935, the Precepts are still prevalent, though with modifications due to micro- and macro-environmental factors. They include "contribute to the development and welfare of the country" and "Be reverent and show gratitude". For example, in his own words in 2011, Mr. Akio Toyoda (President, Toyota Motor Corporation—Japan), during his tour of the devastated parts of Japan following The Great East Japan Earthquake, commented:

> "... I personally went to the scene of the disaster, not only to see for myself the scale of the destruction, but also to listen to voices of the people there. From this experience, I strongly feel that assistance for reconstruction and restoration from the disaster must be on-going, and I instigated the 'Kokoro Hakobu Project' in which our belief that automobiles carry not only goods and supplies, but they can also carry human hearts, is reflected, (Toyota, 2011)".

The CSR approach by CFAO MOTORS UGANDA also aligns with the Toyoda Precepts established in 1935. For example, the 2008–2011 CSR activities were reflective of the personal initiative of the Customer Relations Manager (who also manages CSR) who indicated that:

> As a mother and as a staff [member] of CFAO MOTORS UGANDA, I personally feel concerned and get sleepless nights when I see innocent children suffering. I see them over the news crying most of the days, my soul gets touched. I can't imagine this! Are people so desperate that they even dump innocent babies? What will the future of these babies be?

> [...] Actually, whereas our company's CSR priorities in the last few years were not focused on children directly, in 2011, I decided to celebrate my 37th birthday with children at Sanyu Babies Home [a home for abandoned babies in Uganda]. However, I will ensure in our companies' next CSR strategic planning, children will take a centre stage.

Similarly, during the rebrand of Toyota Uganda, into CFAO in February 2022, the CFAO MOTORS UGANDA Managing Director reiterated:

> "... We have been known as Toyota Uganda since 2005. However, increasingly, we have felt the name does not encompass the full breadth of mobility solutions that we offer, over and above the Toyota brand. The rebrand to CFAO Motors Uganda allows us to align with our CFAO Group philosophy that seeks to strengthen our investment in Toyota whilst at the same time offering additional mobility solutions. More than a name change, this is also a milestone that brings us closer to our ambition: sustainable growth, value chain integration, partnership with strong brands and innovation, consolidating our position as a leading automotive brand" stated Mr. Thomas Pelletier, Managing Director of CFAO Motors Uganda.

Hence, pattern matching revealed an alignment of CFAO Motors (formerly Toyota Uganda), with Toyoda Precepts at corporate level, which ties clearly with the CSR Manager of CFAO MOTORS UGANDA at personal level.

(ii) *Unilever Plc*[9] *and Unilever Uganda Ltd*

- *CSR activities*

Unilever Plc is challenged daily by the resultant effects of poor hygiene, poor nutrition, water scarcity, and climate change. To address these, the Unilever (2011) revealed that, globally, Unilever's CSR activities centre on:

(a) Improving health and wellbeing: Unilever estimates that the company has helped 135 million people worldwide to take action to improve their health and wellbeing. It has focused on health and hygiene (wherein it has reached 100 million people with its handwashing, oral-care, and self-esteem programmes, and a further 35 million with safe drinking water from Pureit); nutrition (wherein it had increased the proportion of its portfolio that meets the highest nutritional standards, from 22% in 2010 to 25% in 2011); and greenhouse gases (where its GHG footprint has remained broadly unchanged).
(b) Reducing environmental impact: As its business grows, Unilever aims to halve the environmental footprint of the making and use of its products. It focuses on water (wherein, come 2020, it will have halved the water consumers use when associating with its products), and waste (wherein, come 2020, the waste associated with Unilever products will have halved).
(c) Enhancing livelihoods: This relates to sustainable sourcing and better livelihoods (contributing to better incomes and livelihoods for farmers and their workers, maintaining soil fertility, enhancing water quality, and reducing

---

[9] Unilever plc is a British multinational consumer goods company with headquarters in London. Unilever products include food, condiments, ice cream, cleaning agents, beauty products, and personal care. Unilever is the largest producer of soap in the world. Unilever's products are available in around 190 countries.

greenhouse gas emissions and protecting biodiversity). In Uganda, for example, on 23 June 2010, Unilever Uganda Ltd., Living Earth Uganda, and German Development Service—DED (*currently* German International Cooperation—GIZ) launched the Plastic Waste Management project in Kampala, which is estimated to have cost approximately USD 40,000 and is aimed at improving sanitation by reducing waste plastic and creating job opportunities for the urban poor.

- *Humanism within the above CSR activities*

Globally,[10] Unilever's CSR activities (highlighted above) are rooted in "Unilever's Five Levers for Change".[11] This is a set of principles developed by Unilever many years ago, following Mr. William Lever's (Unilever's co-founder) moral commitment to a cleaner and healthier environment, and to inspire people to turn their concerns about sustainability into positive actions. Unilever's CSR activities are a clear revelation of Lever's moral commitment to a cleaner and healthier environment, which saw his invention of Lifebuoy soap. The personal interview of the current global CEO of Unilever, Mr. Paul Polman (as reflected in Unilever, 2011) also indicated:

> Unilever has a long history of doing well by doing good. William Lever, one of its founders, created Lifebuoy soap to encourage cleanliness and reduce infectious diseases in Victorian Britain. Today, in the developing world, 2 million people under 5 years die from diarrhoea and respiratory infections daily. Teaching children to wash their hands is a way of reducing this toll. The company sees opportunities to save lives and sells soap [...] We have been working hard to distil those critical areas [Lever's Five Forces] of behavior change insight that we all need to use to engage consumers [...] because we think that there are wider benefits from sharing our work with others.

The company's current CSR activities, coupled with Polman's comment above, reveal that William Lever's moral commitment to a cleaner and healthier environment has been maintained throughout the 100 years in which Unilever[12] has been in business. It also shows how the choice of Unilever's CSR activities is made, as well as how it is shared (by consumers, stakeholders, and all its subsidiaries in developing and developed countries).

(iii) *Bewley's Coffee and Tea Company*[13]

---

[10] Planet & Society | Unilever, https://www.unilever.com/planet-and-society/

[11] Unilever's "Five Levers for Change" are published in a booklet which also contains a series of personal perspectives on sustainable living by leading experts on sustainability and behaviour change from around the world, including Forum for the Future Founder Jonathon Porritt, Akatu Institute for Conscious Consumption President Helio Mattar, and behaviour change specialist Val Curtis from the London School of Hygiene and Tropical Medicine. See http://www.unilever.com/mediacentre/pressreleases/2011/five-levers-change-111123.aspx.

[12] Unilever Sustainable Living Plan, Progress report, 2011 is accessible at: http://www.unilever.com/images/uslp-Unilever_Sustainable_Living_Plan_Progress_Report_2011_tcm13-284779.pdf.

[13] Founded in 1840, Bewley's is Ireland's leading coffee and tea company, after in 1835, Samuel Bewley and his son Charles dared to break the East India Company's monopoly by importing 2099

- *CSR activities*

Bewley's website (2022)[14] and sustainability reports revealed its CSR activities to be centred on: Access scholarship programmes, ethical commitments to business, sustainable trading, environmental protection, and community-based initiatives. Specifically, for over 17 years, Bewley's has been sponsoring the Irish Hospice Foundation's fundraising event, Ireland's Biggest Coffee Morning, Bewley's provides free coffee for this morning event, and each coffee cup is then sold for over €2. The money raised goes to the Hospice Foundation's activities. In addition, since 1995, Bewley's has sponsored the Junior Achievement, an initiative intended to reach out to over seven million young people worldwide. The initiative is aimed at creating a culture of enterprise and a business mindset within the education system, right from primary school through to secondary school. The initiative aims to teach children how they can impact the world around them as individuals, workers, and consumers, and prepares students for their future careers. In addition, for the last 20 years, Bewley's has supported the Dublin City University (DCU) Education Trust Programme (an initiative aimed at providing talented students with the opportunity of a third-level education, which they would otherwise not have attained). In 2007, Bewley's Access Scholarship Programme was also started, with the overall aim of ensuring the sustainability of the local workforce and to further enhance the company's reputation as an employer of choice to future graduates.

In Uganda since 2004, Bewley's has run and supported "*The Traidlinks Initiative*". This is a not-for-profit organization through which Bewley's mentors Ugandans in ethical trading practices so as to be competitive in the food sector (though primary focus is on coffee and tea business players). Broadly, this initiative is meant to link the Irish business community and the Irish Aid programme in developing countries, though with an initial focus on the food sector. Traidlinks' objectives are to promote the relief of poverty and the advancement of education in developing countries. Broadly, Traidlinks Initiative website (2012)[15] provides the following activities being conducted in Uganda, and elsewhere in developing countries:

(a) Encouraging and facilitating the creation of employment opportunities and the improvement of working and living conditions for individuals and communities

---

chests of tea on board the clipper ship. Now counting over 180 years in existence, Bewley's has grown to become one of the largest coffee roasters in the UK and Ireland with an annual turnover more than €170 m. It is the leader in the provision of tea and coffee solutions to the foodservice and retail sectors. Additionally, its award-winning range of teas is supplemented with a wide range of espresso blends and single origin coffees, roasted to suit the varying tastes of today's sophisticated coffee consumers across the globe in Ireland, the United Kingdom, and the USA

[14] Our Company—Bewley's Ireland (bewleys.com), https://www.bewleys.com/ie/about-us/our-company/

[15] http://bewleys.com/about-us/our-ethical-commitment/fairtraide-sustainability/our-initiatives/the-traidlinks-initiative

in developing countries through the development and promotion of trade and business in those countries.
(b) Providing and facilitating the provision of skills exchange programmes, partnering arrangements, operational, managerial, academic, and mentoring schemes in developing countries.
(c) Providing educational opportunities for individuals, communities, and organizations in developing countries for the exchange of knowledge, skills, expertise, information, research materials, advice, and ideas for the development of business, trade, and commerce.
(d) Raising awareness of the role of trade in the development of developing countries.

- *Humanism within the above CSR activities*

To date, Patrick Bewley (the current owner and Chief Executive) has carried on the tradition of his forefathers. That is, a tradition of seeking and engaging in possible ways in which to contribute to the communities within which Bewley's business operations are based. In his own words during the 2010 Ireland's Biggest Coffee Morning, Patrick remarked:

> "… We're delighted to be involved with Ireland's Biggest Coffee Morning, celebrating its seventeenth year in 2010. As in previous years, over a million cups of Bewley's coffee will be served on this special day in September. We want as many people as possible to take part all over Ireland. The more cups of coffee that we can give away, the greater the benefit that will be achieved for hospices and communities all over Ireland".

Bewley's CSR activities have roots in the family history of Mr. Bewley and his son, Charles, who fought to break monopolies in business and trading to level trading grounds as well as ensuring inclusive business. To cement this, currently, the Bewley's company website (2022) states that:

> Corporate Social Responsibility is often presented as a recent development, but the Bewley family has always lived by the principle of social equality. As an Irish, family-owned company, we seek to contribute to the communities within which our operations are based.

This case revealed that the overseas CSR activities at Bewley's are largely skewed towards CSI, while in Ireland a balance is observed between corporate philanthropy, CSI, and Strategic Business Engagement (SBE).

(iv) *Spear Group of Companies Uganda*

- *CSR activities*

Spear Group of Companies Uganda, established in 1974 by Prof. Gordon Wavamunno, with one travel company, is now made up of a chain of companies that includes: the defunct WBS Television, Wavah Water Ltd., Spear Motors Ltd.[16] (a

[16] Spear Motors Ltd. is the Sole official distributor for: (1) Daimler AG of Stuttgart, Germany, the manufacturers of Mercedes Benz cars, & trucks and Mitsubishi FUSO trucks. (2) Fiat Chrysler Automobiles (FCA), the manufacturers of Jeep and Fiat vehicles amongst others. http://www.spearmotors.co.ug/

franchise of Germany's Daimler Chrysler), GM TUMPECO, etc. Key CSR activities of Spear Group are: (i) fosters prisoners' (in-mates') welfare, especially by donating food supplies, health care products, etc.; (ii) facilitates skills development of young professionals (apprentices) where the companies allow graduating students to practise and gain hands-on experience in engineering, metal fabrication, carpentry, etc.; (iii) provide philanthropic support with respect to health and wellness (especially supporting activities of St John's Ambulance); (iv) encourage children's development where the companies, especially WBS TV (*before it was dissolved*) and Wavah Water, contribute generously to programmes aimed at children's up-bringing; and, lastly (v) promoting arts and culture and upcoming music artists in Uganda, especially through the Pearl of Africa Music Awards (PAM), which are organized annually in Uganda to recognize and promote music as a talent and as a way of living.

Unlike other companies studied, the CSR activities of Spear Group of Companies are largely philanthropic in nature. There are limited traces of CSI (for instance, in apprenticeship) and strategic business engagement. This could be a result of the group's infancy in CSR activities when compared to the other companies studied but could also be an entry point into other dominant forms of CSR engagement.

- *Humanism within the above CSR activities*

The founder of this chain of businesses, Prof. Gordon Wavamunno,[17] was found to be behind the CSR agenda and decisions of his portfolio. His discretion and business power have directed his business portfolio choice and selection of CSR activities. Describing his decision to engage in such CSR activities, he revealed that the political history of Uganda and the economic upheavals prompted and compelled him to meet certain obligations, and indeed be committed to ensure tangible results emerge from such engagements. In his own words, Prof. Gordon Wavamunno, while reacting to one of his companies winning the CSR Exemplary Award in Uganda, commented:

> "… We are not competing with anybody in terms of helping anybody. We just offer to the public what we think we must. It is not surplus that we have and that surplus we should give out…"

[17] Prof. Gordon Babala Kasibante Wavamunno, was born on 16 December 1943. He is a Ugandan entrepreneur, businessman, and philanthropist. He is the owner of Spear Motors Uganda [a franchise/sole distributor of Mercedes Benz in Uganda]. He was the director of the defunct WBS Television, and is counted amongst the wealthiest people in Uganda. Lastly, he serves as the honorary consul of Hungary to Uganda.

Discussion

CSR activities of companies (especially those with operations in Uganda) have grown to their current stance mainly because the individuals (Samuel Bewley, Kichiro Toyoda, William Lever, and Prof. Gordon Wavamunno) were behind their organizations' CSR practices. Hence, from the findings, the following humanism themes emerged:

(i) *Sensuality*

We observed that the five human senses (sight, smell, taste, touch, and hearing) attributed to Aristotle (McLeisch, 1999), guide human beings in managing life, including organizational matters. Toyota's Toyoda Precepts, Bewley's sense of inclusive business, as well as William Lever's sense of belonging to a cleaner and healthier environment, revealed a close match and sharing of sensuality. These cases indicated that the functionality of human beings' senses plays a big role in triggering an individual's understanding, thereby forming, and guiding his/her organization's CSR direction and choice. For example, an organization's individual (staff, owner, board member, etc.) is triggered into action if he/she hears (e.g. from media, seminars, etc.), sees (e.g. the unethical trading practices in coffee and tea business witnessed by Mr. Bewley and his son Charles; earthquake catastrophe which Mr. Akio Toyoda saw in Japan, etc.), tastes (e.g. adulterated products in the markets such as coffee and tea for edibles, or mechanical components for motor vehicle consumption), or smells (e.g. the filthy environment or polluted air that can cause disease like pneumonia, as witnessed by Mr. William Lever). The findings, however, revealed that sight is the most developed sense in humans when it comes to CSR. These findings are supported by philosophers Aristotle (McLeisch, 1999) and Immanuel Kant (Hartnack, 2001; Wood, 1984, 2005), whose works argue that human beings' knowledge of the world around them, *vis-à-vis* their response(s), depends on the functionality of their senses. Kant profoundly indicated that human beings (in this case, business owners, founders, and/or managers) cannot easily just sit back (Wood, 2005, p. 84) when a CSR-worthy activity appears. They will be aroused to devise responsive CSR actions.

(ii) *Family/personal history*

Historical experiences, life, interests, and the spirit of the company founders and mangers play a central role in determining a company's CSR orientation, the nature and choice of CSR activities they decide to become engaged in. For example, the Toyodas' (Sakichi, Kichiiro, and Akio) experience of how their great grandmother suffered with the inefficient loom, along with the interests and history of Sakichi regarding innovations and inventions, determined—and still guide–Toyota's CSR orientation. Similarly, Unilever's CSR activities are shaped from William Lever's experience with problems relating to poor hygiene and sanitation. At the same time, the unethical trading practices (especially monopolistic tendencies of merchants) that Mr. Bewley and his son Charles witnessed, have guided Bewley's CSR to embrace ethical trading practices that include mentoring farmers to become responsible merchants, as well as offering scholarships in the fields of agriculture and trade. We observed that these historical experiences, life, and spirit have continuously been passed on to new generations of business owners and managers through established channels of communications, values, and norms. Furthermore, findings proved that as a company grows over the years, the person (current business owner or manager in charge of CSR) naturally has it at his or her heart to build on the (family) history of his/her forefathers and to ensure their reflection in the company's operations. The CSR activities of companies with this theme (family/personal history) displayed a reasonable proportion of maintaining their CSR engagement in philanthropic activities. This method of preserving and passing on experiences confirms agency theory (Jensen & Meckling, 1976) and the works of Eisenhardt (1989), who indicated that much of organizational life is based on self-interest and the experiences of individuals. The findings also confirm the views of Katamba and Nkiko (2016) who disclosed how CSR transcend through historical values of the society to current behavioural trend. Lastly, this preservation method supports works of Lieberman's (1979) trans-generation theory, which indicated that unless history and values are passed through a business from generation to generation, their meaning and value to business will be lost along the way.

(iii) *Moral commitment*

From the companies studied, we found that business owners, founders, and/or current managers, have a moral commitment to transform their surroundings. The activities that followed this pattern were largely CSI-related in nature, and to a certain extent entailed strategic business engagement. For example, William Lever had a moral commitment to a clean and healthy environment. Similarly, Samuel, Charles, and Patrick Bewley have a moral commitment to ethical trading practices which have seen Bewley's Coffee and Tea Company expanding across borders. However, because of business legal framework (corporate veil) that provides for separation of personality of individuals from corporations, we found that they feel their moral commitment can best be enacted through their companies, trusts, trustees, foundations, or other institutions. For example, William Lever's "moral commitment" saw him inventing Lifebuoy soap, which is now a star product for Unilever. This means that if their companies do not pursue these commitments/responsibilities, the individuals feel cheated. The findings also revealed that moral commitment makes human beings act with intention, aim to achieve goals, and be aware that they have an impact on future events. For example, Unilever's Five Forces of Change, the Toyoda Precepts, and Bewley's ethical traditions have contributed much to forming strong and sustainable foundations for their companies' social responsibilities. Interestingly, we cross-examined findings in this theme: "moral commitment" against psychologist Maslow's (1954). Hierarchy Triangle and Bugental's (1964) third force in psychology, as well as Baier's (1993) egoism and Greening's (2006) postulates of human psychology. The revelation was that there is a natural fit between moral commitment and self-actualization, achievement, love and belonging, safety needs, and the basic needs of individual businessmen. However, this natural fit is attributed to companies in the form of distinguished reputation, unmatched competitiveness, risk mitigation, and opportunity maximization. This, therefore, reveals that corporate CSR activities are, in essence, a reflection of humanism. Their sustainability even after their [CSR] original proponents (individual businessmen) have passed on, signifies actualization and achievement for the proponents, though attributed to companies.

Institutionalizing CSR Guided by Humanism of CSR

From the findings and discussion above, we present a summarized model for the humanism of CSR activities, which we call "THE HUMANISM OF CSR MODEL" as shown in Fig. 2.1 below. This model is a tool intended to guide founders, owners, and/or persons in charge of CSR activities to understand humanistic [human driving] factors behind the CSR activities they undertake. This is a key issue if CSR is to be sustainably institutionalized.

THE HUMANISM OF CSR MODEL attempts to demonstrate that the CSR activities we see organizations undertaking, are a reflection, transposition, and modification of intrinsic factors (sensuality, family background, personal history, and moral commitment) of the human being(s) behind the organization, or its CSR activities. However, these are subject to extrinsic motives such as stakeholder activism, international standardization, supply chain requirements, and legal requirements (Visser, 2008a, p. 481). In essence, the model considers these under testers, filters, or modifiers. That is, business must go through modifications like mergers, acquisitions, new product developments, or product line extensions, or even are subjected to passing a test of time (perpetual) brought about by different generations, economic situations or revolutions, and family generation changes. The best

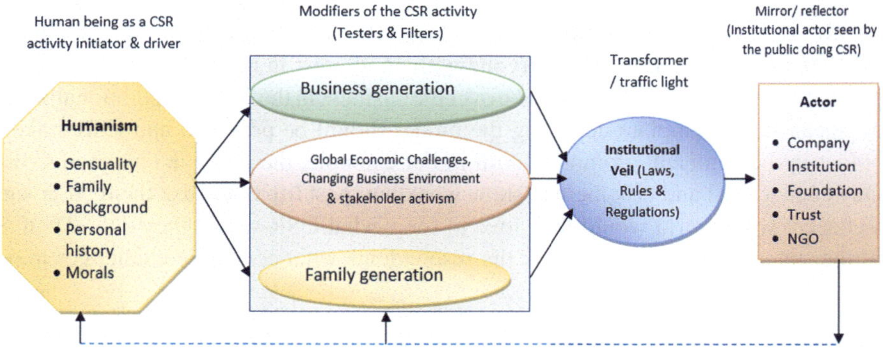

Fig. 2.1 The humanism of CSR model. Source: Developed after study findings and discussions

example here are the mergers and acquisitions which transformed Mercedes Benz[18] (*Adler,* 2006*; Daimler,* 2012*), and Toyota Uganda Ltd becoming CFAO Motors Uganda Ltd*[19] *in 2022.* This means that through all these tests, filters, or modifiers, a need to safeguard the humanistic intentions of the "*CSR activity initiator*" arises. This need may then be aligned to what the prevailing business laws provide, say by either registering or forming a company, foundation, trust, project, or institution, or a combination of these, which will be taken as different from the persons running it (Institution/corporate veil). This model, therefore, has identified humanism as being subjective and/or changing over generations, and thus the CSR activities must follow the same trend. Therefore, due to humanism's subjectivity to extrinsic drivers, CSR activities of companies cannot be a "direct reflection" of the human(s)/initiator(s) of CSR activities, but a transformed reflection. This justifies the modest/slight difference observed in the choice of CSR activities/undertakings of companies.

- *Recruiting CSR activities managers*

From the model, it becomes apparent that recruiting someone to oversee CSR activities of organizations should go beyond merely focusing on the academic qualifications and experiences of the candidate. Behavioural, psychological, and historical assessments into how the person's life evolved in relation to CSR-related aspects should be investigated. For example, by introducing ethical/sustainability questions into the psychometric framework of the company or its recruitment agency; asking a straightforward question with the aim of digging into the candidate's background, with the interviewer looking for a life-history story/act along the lines of CSR or sustainability conduct. This will help to identify a creative and committed person or one who shares the CSR aspirations of the recruiting company.

- *Prepare the next generation to assume the CSR of companies*

For companies/organizations that are lucky enough for their proprietors to be living and/or that can provide wisdom to the CSR direction of the business/organization, efforts should be instituted and maintained that work towards the transferral of their CSR orientation to the next generation of managers. This can be through openly sharing and operationalizing "THE HUMANISM OF CSR MODEL" FRAMEWORK/MODEL for everyone with an interest in the company; for example, staff or even those lined up to take over the business in the future. Operationalizing we mean that a template depicting the model should be provided and CSR stakeholder in the company prompted to fill it. For example, the office in charge of CSR may engage staff and task them to show what their past life is and ask them "if given a chance, what world would they love to see—what CSR would they do?" Then a further interaction goes on to task the respondents what economic challenges may

[18] The merger between Daimler-Benz and Chrysler to DaimlerChrysler AG (1995–2007) | Mercedes-Benz Group > Company > Tradition > Company History, https://group.mercedes-benz.com/company/tradition/company-history/1995-2007.html

[19] Toyota Uganda Limited becomes CFAO Motors Uganda - CFAO (cfaogroup.com), https://www.cfaogroup.com/en/toyota-uganda-limited-becomes-cfao-motors-uganda/

make their CSR aspiration successful or how would they go about it... what stakeholders can they engage with to realize that CSR aspiration. Lastly, they can then be tasked to guide their company how best to implement such a CSR activity.

Conclusions

CSR activities/undertakings in organizations are a clear reflection, transposition, and modification of emotions, values, family/personal history, and spirituality of the human(s) behind such activities or behind the organizations themselves. However, because of the corporate veil and other extrinsic factors, these human beings are limited to act in their individual human capacity, rather than through establishments (foundations, companies, trustees, etc.). These humans include (either collectively or individually) those initiating CSR activities, and manning them, financing them, formulating CSR policies, or subscribing to them (policies and activities).

The study shows the importance of psychological application towards understanding why founders/owners and/or managers engage in certain CSR practices. Evidence put forward relates founders'/owners' engagement in CSR practices towards sensuality, family, personal history, and moral commitment, which confirms that there is a psychological element driving CSR engagement.

The findings also revealed that CSR undertakings of parent companies and their subsidiaries elsewhere in the world entail a trans-generation of the specific human histories, experiences, and moral commitments that drove the formation of these companies or their engagement in CSR-related aspects.

Lastly, from this study's findings and discussion, we conclude with the meaning and interpretation of the humanism of CSR activities we have drawn from attempts that conceptualize, theorize, and marry human nature into the CSR debate. This definition is:

> [Humanism of CSR activities is] the integration, reflection and propelling of a human being's personality, interests, preferences, life, behaviour, and development in the introduction, realization, and sustainability of CSR activities of an organization prior to or even without hope of receiving any return from the recipients and beneficiaries of such CSR activities.

References

Adler, D. (2006). *Daimler and Benz: The complete history – The birth and evolution of the Mercedes-Benz*. HarperCollins. ISBN 0060890266.

Asongu, J. J. (2007). The history of corporate social responsibility. *Journal of Business and Public Policy, 1*(2), 1–18.

Baier, K. (1993). Egoism. In P. Singer (Ed.), *A companion to ethics*. Blackwell.

Baskin, J. (2006). Corporate responsibility in emerging markets. *Journal of Corporate Citizenship, 24*(Winter), 29–47.

Bewley's website. (2022). Available from http://bewleys.com/about-us/our-ethical-commitment/corporate-social-responsibility. Accessed 25 May 2022.
Bowen, H. R. (1953). *Social responsibilities of the businessman*. Harper & Row.
Bugental, J. (1964). The third force in psychology. *Journal of Humanistic Psychology, 4*(1), 19–26.
CAPAM. (2010). *Overview of case study models and methodology*. A Shared Library of Public Administration Case Studies (CAPAM/Commonwealth Secretariat Project). Available from http://www.capam.org/_documents/reportoncasestudymethodologies.pdf Accessed on 25 May 2022.
Carey, B. (2005). *Chancellors, commodores, & coeds: A History of Vanderbilt University*. Clearbrook Press.
Carroll, A. B. (1979). A three-dimensional conceptual model of corporate social performance. *Academy of Management Review, 4*(4), 497–505.
Carroll, A. B. (1999). Corporate social responsibility: Evolution of a definitional construct. *Business and Society, 38*(3), 268–295.
Chernow, R. (1998). *Titan: The Life of John D. Rockefeller, Sr*. Random House Group.
Connelly, J. (2010). T.M. Knox and the study of economic activity. *International Journal of Social Economics, 37*(11), 880–893.
Daimler AG. (2012). *Daimler Chronicle*. Göttingen: Steidl Verlag. ISBN 9783869309835.
Davis, K. (1967). Understanding the social responsibility puzzle. *Business Horizons, 10*(4), 45–51.
EC (European Commission). (2001). *Promoting a European framework for corporate social responsibility*. Green Paper, European Commission, Directorate-General for Employment and Social Affairs, Brussels, July 2001.
Eisenhardt, K. M. (1989). Agency theory: An assessment and review. *The Academy of Management Review, 14*(1), 57–74.
Follett, M. P. (1896). *The speaker of the house of representatives*. Longman Green.
Follett, M. P. (1924). *Creative experience* (reprinted by Peter Owen in 1951). Longman Green, New York.
Follett, M. P. (1927). *Dynamic administration* (reprinted in 1942). Harper & Brothers Publishers, New York.
Fombrun, C. J., & Van Riel, C. B. M. (2004). *Fame and fortune: How successful companies build winning reputations*. FT Prentice-Hall.
Friedman, M. (1970). The social responsibility of business is to increase its profits. *The New York Times Magazine, 13 Sept*, 122–126.
Gilbreth, L. M. (1914). *The psychology of management: The function of the mind in determining, teaching and installing methods of least waste*. Sturgis and Walton.
Graham, L. (2000). Lillian Gilbreth and the mental revolution at Macy's, 1925–1928. *Journal of Management History, 6*(7), 285–300.
Grayson, D., & Hodges, A. (2004). *Corporate social opportunity! 7 steps to make corporate social responsibility work for your business*. Greenleaf Publishing.
Greening, T. (2006). Five basic postulates of humanistic psychology. *Journal of Humanistic Psychology, 46No. 3*, 239–239.
Hartnack, J. (2001). *Kant's theory of knowledge: An introduction to the critique of pure reason*. Hackett Publishing Company.
Heald, M. (1957). Management's responsibility to society: The growth of an idea. *Business History Review, 31*(4), 375–384.
Heald, M. (1970). *The social responsibilities of business: Company and community, 1900–1960*. The Press of Case Western Reserve University.
Hemingway, C. A., & Maclagan, P. W. (2004). Managers' personal values as drivers of corporate social responsibility. *Journal of Business Ethics, 50*, 33–44.
Idowu, S. O. (Ed.). (2021). *Current global practices of corporate social responsibility. CSR, sustainability, ethics & governance*. Springer Nature, Switzerland.
Jensen, M., & Meckling, W. (1976). Theory of the firm: Managerial behavior, agency costs and ownership structure. *Journal of Financial Economics, 3*, 305–360.

Johansson, R. (2003). *Case study methodology.* Key note speech given at the International Conference Methodologies in Housing Research, organized by the Royal Institute of Technology in cooperation with the International Association of People–Environment Studies, Stockholm, 22–24 September 2003.

Katamba, D., Kazooba, C. T., Mpisi, S. B., Nkiko, C. M., Nabatanzi-Muyimba, A. K., & Kekaramu, J. H. (2012). Corporate social responsibility management in Uganda: Lessons, challenges, and policy implications. *International Journal of Social Economics, 39*(6), 375–390.

Katamba, D., & Kekaramu, J. H. (2012). Getting involved in corporate social responsibility. In D. Katamba, C. Zipfel, C. Tusahabomwe-Kazooba, & D. Haag (Eds.), *Principles of Corporate Social Responsibility: A guide for students and practicing managers in developing and emerging markets.* Strategic Book Publishing and Rights Agency.

Katamba, D., & Nkiko, C. M. (2016). The landscape of corporate social responsibility in Uganda: Its past, present and future. In S. Vertigans, S. O. Idowu, & R. Schmidpeter (Eds.), *Corporate social responsibility in sub-Saharan Africa: Sustainable development in its embryonic form* (pp. 153–171). Springer. http://link.springer.com/chapter/10.1007%2F978-3-319-26668-8_7

Katamba, D., & Wickert, C. M. J. (2021). Corporate social responsibility in Uganda. In S. O. Idowu (Ed.), *Current global practices of corporate social responsibility. CSR, sustainability, ethics & governance* (pp. 563–578). Springer. https://doi.org/10.1007/978-3-030-68386-3_26

Katamba, D., Zipfel, C., Tusahabomwe-Kazooba, C., & Haag, D. (2012). *Principles of Corporate Social Responsibility: A guide for students and practicing managers in developing and emerging markets.* Strategic Book Publishing and Rights Agency.

Keim, G. D. (1978). Corporate social responsibility: An assessment of the enlightened self-interest model. *Academy of Management Review, 3*(1), 32–39.

Layder, D. (1998). *Sociological Practice. Linking theory and social research.* Sage.

Levy, S. (1988). *Information technologies in universities: An institutional case study.* Unpublished doctoral dissertation. Northern Arizona University, Flagstaff.

Lieberman, S. (1979). A trans-generational theory. *Journal of Family Therapy, 1*(3), 347–360.

Maslow, A. (1954). *Motivation and personality.* Harper and Row.

McLeisch, K. C. (1999). *Aristotle: The great philosophers.* Routledge.

McLeod, S. A. (2008). *Simply psychology; case study method in psychology.* Available from http://www.simplypsychology.org/case-study.html (accessed June 9, 2022).

Melé, D., & Castella, J. M. (2010). *El DesarrolloHumano integral.* Iter.

Melé, D., & Mammoser, T. L. (2011). Humanistic corporate community involvement: Walgreens/MEF One-On-One program. *Journal of Management Development, 30*(6), 582–593.

Miles, M., & Huberman, M. (1984). *Qualitative data analysis: A source book for new methods.* Sage.

Nkiko, C. M. (2009). *Corporate Social Responsibility: Small medium enterprise contribution towards sustainable development in developing economies: The case of BCSDU.* Unpublished DBA thesis, University of Portsmouth.

PCJP (Pontifical Council for Justice and Peace). (2004). *Compendium of the Social Doctrine of the Church [WWW].* LibreriaEditriceVaticana, Vatican. Available from www.vatican.va/roman_curia/pontifical_councils/justpeace/documents/rc_pc_justpeace_doc_20060526_compendio-dott-soc_en.html (Accessed May 15, 2022).

Pope John Paul II. (1991). *Letter encyclical 'centesimusannus'.* Available from www.vatican.va/edocs/ENG0214/_INDEX.HTM (Accessed June 10, 2022).

Porter, M., & Kramer, M. (2002). The competitive advantage of corporate philanthropy. *Harvard Business Review, 80*(12), 57–68.

Porter, M., & Kramer, M. (2006). Strategy and society: The link between competitive advantage & corporate social responsibility. *Harvard Business Review, December, 2006*, 78–92.

Ritchie, R., & Mispy, O. (2018). *Measuring progress towards the Sustainable Development Goals.* SDG-Tracker.org, website (2018).

Sjoberg, G., Williams, N., Vaughan, T., & Sjober, A. (1991). The case study approach in social research. In J. Feagi, A. Orum, & G. Sjoberg (Eds.), *A Case for Case study*. University of North Carolina Press.

Stake, R. (1995). *The art of case research*. Sage.

Stake, R. (2000). Case studies. In N. Denzin & Y. Lincoln (Eds.), *The handbook of qualitative research* (pp. 435–454). Sage.

Stiles, T. J. (2009). *The First Tycoon: The Epic Life of Cornelius Vanderbilt*. Vintage.

Tellis, W. (1997). Information technology in a university: A case study. *Campus-Wide Information Systems, 14*(3), 78–91.

Toyota. (2011). *Citizenship report*. Available from http://www.toyota-global.com/sustainability/report/citizenship/pdf/citizenship.pdf. Accessed May 25, 2022.

Traidlinks Initiative website. (2012). Available from http://www.traidlinks.ie/activities/bus-mentorprog. Accessed 25 May 2022.

Unilever. (2011). *Sustainable living plan progress report*. Accessible from http://www.unilever.com/images/uslp-Unilever_Sustainable_Living_Plan_Progress_Report_2011_tcm13-284779.pdf. Accessed May 25, 2022.

United Nations Economic and Social Council. (2020). *Progress towards the Sustainable Development Goals Report of the Secretary-General*, High-level political forum on sustainable development, convened under the auspices of the Economic and Social Council (E/2020/57), 28 April 2020.

Vatican Council II. (1965). *Pastoral constitution: Gaudiumetspes*. Available from www.vatican.va/archive/hist_councils/ii_vatican_council/documents/vat-ii_const_19651207_gaudiumet-spes_en.html (Accessed May 15, 2022).

Visser, W. (2008a). *Making a difference: Purpose-inspired leadership for corporate sustainability & responsibility*. VDM.

Visser, W. (2008b). Corporate social responsibility (CSR) in developing countries. In A. Crane, D. Matten, A. McWilliams, J. Moon, & D. Siegel (Eds.), *The Oxford handbook of corporate social responsibility*. Oxford University Press.

White, L. A. (2006). *Business brief: Intangibles and CSR, Business for social responsibility*, February, pp. 6.

Widdersheim, M. M. (2018). Historical case study: A research strategy for diachronic analysis. *Library & Information Science Research, 40*(2), 144–152. https://doi.org/10.1016/j.lisr.2018.06.003

Wood, A. W. (Ed.). (1984). *Self and Nature in Kant's philosophy*. Cornell University Press.

Wood, A. W. (2005). *Kant*. Wiley.

World Bank Institute. (2002). *E-Course on corporate social responsibility and sustainable competitiveness*. World Bank.

World Business Council for Sustainable Development (WBCSD). (2000). *Corporate social responsibility, making good business sense*, January, pp. 9.

Wren, D. A. (2005). *The history of management thought* (5th ed.). Wiley.

Yin, R. (1984). *Case study research: Design and methods*. Sage.

Yin, R. (1994). *Case study research: Design and methods*. Sage.

David Katamba, PhD, is an Associate Professor of Corporate Sustainability and Responsibility. I am an expert evaluator with Economic and Social Research Council (ESRC), a part of UK Research and Innovation (UKRI). I initiated and led the consultative process that established the United Nations (UN) Global Compact Network in Uganda, which is currently hosted by Federation of Uganda Employers (FUE). I am a World Bank Institute trained Corporate Sustainability and Responsibility (CSR) professional with over 16 years' experience in Higher Education quality management systems, sustainable development aspects including in youths and women entrepreneurship, Education, and Business. Currently (Jan 2023–June 2026), I am coordinating a €

1,000,000 EU Erasmus+ funded project, 'Strengthening Capacity for ICT Usage and Policy Reforms for Relevant and Quality Masters Research Process Management in Uganda's Higher Education Institutions (ICT-4MRPQ)', (project No. 101083105). He is a certified CSR Consultant/ Advisor and trainer for International Business Leaders Forum (IBLF) and Aga Khan Development Network (AKDN). I have won and led implementation of several international research grants including from €300,000 Netherlands' NWO WOTRO to implement AGRI-QUEST project (Project No. W 08.270.317). I was also part of team that won €801,000 from EU Erasmus+ [call, EAC/A03/2016, research project No: 574218-EPP-1-2016-1-NO-EPPKA2-CBHE-JP to implement CapSEM project (2016–2019). Dr Katamba has supported several HEIs in and outside Uganda to strengthen their academic quality processes through coaching services, matchmaking, financial resources mobilization, and capacity building. I have been (and I am still) a peer reviewer of proposals, grant applications, publications for several academic journals, and projects for several development organizations including for European Commission's EACEA. Additionally, he acted as mentor, coach, and match makers for 'Start Your Business (SYB)' scheme and competitions by World Bank and Private Sector Foundation Uganda (PSFU) as well as other SME and trade and investment initiatives. He has delivered services and implemented projects in Netherlands, Belgium, Canada, Turkey, Italy, and East Africa. Lastly, he helped to start the United Nations (UN) Global Compact Network in Uganda.

Christopher M. J. Wickert, PhD, is Professor of Ethics & Sustainability and Director of the VU Center for Business & Society, Amsterdam. He is also Research Director of the Department of Management & Organization at VU Amsterdam. His research examines corporate social responsibility (CSR) and corporate sustainability as well as the broader relationship between business and society by drawing on various strands of organization and management theory. Christopher obtained his PhD in Management from the University of Lausanne in Switzerland in 2013. During his doctoral studies, he has also conducted research and taught at University of Zurich, University of St. Gallen and at IESE Business School in Barcelona. Prior to becoming a researcher, he worked as a consultant with the United Nations Global Compact Office in New York. He holds an MSc degree in business administration from WHU Otto Beisheim School of Management in Vallendar, Germany. In his research, Christopher explores different patterns of corporate social responsibility implementation in globally integrated companies, paying particular attention to how firm size influences the way companies organize CSR. Furthermore, Christopher has been studying a range of phenomena under the umbrella of CSR, such as the evolvement and diffusion of CSR standards, the influence of organizational identity orientation on the implementation of CSR practices, or the prevalent mismatch between CSR 'talk' and CSR 'walk' in small and large firms. Furthermore, he has investigated dynamics at the individual level of analysis by looking at CSR professionals inside large corporations and their concomitant struggles when selling CSR issues intra-organizationally. Christopher is also interested in Critical Management Studies (CMS), philosophy of science, business ethics, and social entrepreneurship. In 2019, he published a book with Cambridge University Press, 'Corporate Social Responsibility (Elements in Business Strategy)', which provides a critical introduction to contemporary approaches to CSR. Christopher's research has appeared in journals such as Academy of Management Discoveries, Business & Society, Harvard Business Review, Human Relations, Journal of Business Ethics, Journal of Management Studies, and Organization Studies, as well as in several book chapters. He was associate editor of Business & Society and is currently associate editor of the Journal of Management Studies.

Cedric Marvin Nkiko, PhD, is an Associate Professor at Worcester University. He is a strategic and creative leader with an extensive international career at the crossroads of sustainability, management, and strategic leadership and has an established record of creating ambitious academic programmes with proven impact. Besides teaching and researching at Worcester University, Cedric consults for businesses, NGOs, and governments, in areas of leadership strategic corporate responsibility & governance, sustainability strategy & sustainable investment & operations management.

He continues to work on different consultancy projects for clients such as the UNDP, WBCSD, Rolls-Royce, PayPal, NHS, and several local authorities in the UK. Cedric has supported leaders of some of the world's leading companies in China, Japan, Brazil, the USA, Spain, Germany, Switzerland, Australia, Botswana, Nigeria, and Uganda to challenge current practices and help their companies to achieve transformational change. He has a passion for democratizing learning and promoting engaging and accessible educational opportunities for future learners in blended and online formats.

Chapter 3
CSR Disclosure, Motivation for Disclosure, and Who Matters to the Firm: Prospecting for Normative Oraganisational Practice

Uzoechi Nwagbara, Yahaya Alhassan, Ngozi Ibeawuchi, and Jacyntha Stewart

Introduction

Corporate social responsibility (CSR) is generally conceived as obligation of firms or organisations to its various stakeholders (including the environment and society) beyond what the law specifies (Crane & Matten, 2010; Visser, 2013). It can also mean a firm's commitment to practise social and environmental sustainability to be deemed by stakeholders as a good corporate citizen in its operating environment (Crowther & Aras, 2008). CSR can be appropriated by organisations to court legitimacy (Idemudia, 2010; Deegan, 2002), decrease stakeholders' criticism (Belal, 2008), increase level of trust (Ferguson et al., 2016), manage reputation (Nwagbara & Belal, 2019), and increase "social licence" to operate (Higgins & Walker, 2012). For a long time, organisations have engaged in CSR and its reporting, yet one question that persists is the impact of their CSR initiatives/activities on the stakeholders and their environment (Ferguson et al., 2016; Ite, 2004). Therefore, CSR has remained a problematic and contested concept in developing countries, specifically (Okoye, 2009).

For instance, Zwetsloot (2003) contends that CSR (reporting) appears to be a leading principle for business reinvention, innovation, and sustainability aimed at

U. Nwagbara · Y. Alhassan (✉)
University of Sunderland in London, London, UK
e-mail: uzoechi.nwagbara@sunderland.ac.uk; Yahaya.alhassan@sunderland.ac.uk

N. Ibeawuchi
Coventry University London, London, UK
e-mail: Ngozi.ibeawuchi@coventry.ac.uk

J. Stewart
For Business Sake, London, UK
e-mail: info@forbusinesssake.org

U. Nwagbara et al. (eds.), *Corporate Social Responsibility Disclosure in Developing and Emerging Economies*, CSR, Sustainability, Ethics & Governance, https://doi.org/10.1007/978-3-031-61976-2_3

reinventing organisational practices that can impact positively on the environment, profit, and people (Crowther & Aras, 2008). Accordingly, Emeseh and Songi (2014) argue that CSR reporting is central to business responsibility, transparency (Visser, 2013). Notwithstanding global effort at encouraging organisations to engage in CSR, research demonstrates (see Frynas, 2005) that the processes through which they do so are undermined by managerial ideology and/or orientation, eliciting stakeholders' criticism (Okoye, 2009). Consequently, stakeholders have contended that managerial orientation is mainly responsible for apparent lack of social responsibility (Okoye, 2009), transparency (Crane & Matten, 2010), accountability (Belal, 2008), legitimacy (Deegan, 2002), and neglect of stakeholder engagement (Idemudia, 2010; Disu & Gray, 1998; Gray, 2002; Ite, 2004; Nwagbara & Belal, 2019).

Furthermore, although approaches to CSR vary, some organisations engage in CSR as a means to advance shareholders' interest (Freeman, 1984), increase profit (Belal, 2004, 2008; Freeman, 1984), maintain legitimacy (Nwagbara & Belal, 2019), and strategically position their operations (Crane & Matten, 2010; Eweje, 2007); while some organisations engage in CSR initiatives out of a sense of moral obligation to society (Emeseh & Songi, 2014) and the environment (Elkington, 1997). Given that CSR is a voluntary initiative and/or activity (Visser, 2013), firms tend not to be accountable, transparent, and responsible in their CSR activities (Deegan, 2002). Nevertheless, this situation is more complex in developing countries, such as Nigeria, where there are glaring failure of organisations, specifically oil multinational corporations, in being responsible in their operations (Idemudia, 2010), instances of weak institutions (Nwagbara & Alhasan, 2021, poor governance regime (Frynas, 2005), issue of "resource curse" (Karl, 1997), paradox of plenty (Ross, 1999), and political leadership challenges (Alazzani et al., 2019), which have impact of CSR and how it is reported (Nwagbara & Alhasan, 2021; Belal, 2008). Taken together, the preceding argument signals who matters to the firm as business is traditionally conceptualised and established to make profit for the shareholders, while leaving the rest of the wider stakeholders at the mercy of the financiers of business (Freeman, 1984). This argument echoes the idea of "business in society" as opposed to "business and society"; the former is about business helping in healing social and environmental wounds, while the latter is about profitability and shareholder value maximisation even within the remit if CSR reporting (Nwagbara & Belal, 2019).

From CSR to CSR Reporting: CSR Reporting and Rationale for Disclosure

CSR reporting is conceptualised as corporate response to stakeholder claims (Alazzani et al., 2019; Belal, 2008). It generally entails deliberate, timely, and formal disclosure of a firm's CSR engagement and commitment in its operating environment as well as the impacts its operations have on the people and environment (Carroll & Shabana, 2010). Given that CSR reports, which are not regulated by law (Idemudia & Ite, 2006; Nwagbara & Belal, 2019), are prepared by organisations, firms have a tendency to position themselves as good corporate citizens; they can use these reports for propaganda in the way they report their footprints on society and the environment (Frynas, 2005); or use them to engage in impression management (Scherer & Palazzo, 2011) as well as persuade wider stakeholders that they are responsible in their reporting (Nwagbara & Belal, 2019), which has attracted local and international condemnations (see important reports like UNEP, 2011 and Amnesty., 2015). Various forms of corporate reporting/disclosures have developed lately including sustainability reports, press releases, annual reports, social and environmental reports, and the like delineating several social and environmental initiatives, which are all referred to as CSR disclosure (Gray et al., 1996). As observed by Chantziaras et al. (2020), CSR reports enable organisations, to respond to wider stakeholders who continually question the accountability and transparency of their activities as well as untrustworthiness of these organisations' actions, business orientation and for the foremost part, motivations for reporting (Nwagbara & Belal, 2019). Besides, CSR reports can broaden the accountability of organisations to wider stakeholders beyond sheer preparation of traditional financial accounts (Egbon & Mgbame, 2020), by so doing interests of other users/stakeholders of information contained in these reports are put into consideration (Chantziaras et al., 2020; Gray et al., 1995). Corporate disclosure, including CSR reporting, has precipitated a considerable amount of research from practitioner and academic viewpoints in the last few decades (Nwagbara & Belal, 2019). Consequently, Nwagbara and Belal (2019) contend that part of the motivation for this rise in interest in CSR disclosure includes its implication in corporate reporting quality, transparency of organisational actions, credibility of organisational behaviours, organisational legitimacy and related reasons (KPMG, 2017; Deegan, 2002; Nwagbara & Belal, 2019). Landmark regulatory reforms including Global Reporting Initiative (GRI) and Sarbanes–Oxley Acts (SOA) were initiated as a consequence (KPMG, 2017).

The Rise of CSR Disclosure: From Traditional Accounting to CSR Disclosure

Given the emphasis of traditional accounting on monetary and financial statements of firms' accounts, which largely serves the interests of shareholders as well as creditors, social accounting rose to "arrest" this development (Blowfield & Murray, 2011; Kamla, 2007). In capitalist, industrial society in order to ensure efficient use of capital, traditional accounting is used in maintaining mechanisms to protect shareholders (for example, restriction on distributable profits) including reporting provisions (for example, publication of annual reports) (Hopper et al., 2009). This state of affairs is generally agreed by most scholars on accounting literature as mediated by the collusion of the state, the establishment, and corporations to perpetually further information asymmetry (Asechimie, 1992), which can be a harbinger of managerial capture.

In taking this perspective further, Elliott and Elliott (2006, p. 622) have argued that the state issues laws to ensure effective control of the capital market; the degree of intervention is hinged on the political party in power. Likewise, accountants issue principles and standards to ensure there is reliable information to shareholders to create and maintain an orderly capital market. Both the state and the accountancy profession have directed their main efforts towards servicing the ideals of capital accumulation. In ensuring public interest accounting as well as in recognising pressures from stakeholders for firms to ensure accountability and comprehensive coverage of both financial and non-financial activities in their disclosures, social accounting was given birth to. In this connection, as Kamla (2007) contends, "the end purpose of social accounting/reporting (CSR reporting) would be about enabling accounting to participate in achieving the welfare of the whole society" (p. 108, my parenthesis).

Social disclosure has a history dating back to the beginning of the twentieth century (see Owen, 2005; Guthrie & Parker, 1989). As a consequence, social disclosure is far from being a contemporary phenomenon. Nevertheless, as Owen and O'Dwyer (2008) assert, social reporting and disclosure "… achieved real prominence in the 1970s, largely as a consequence of the debate then raging concerning the role of the corporation in society at a time of rising social expectations and emerging environmental awareness" (p. 386). However, as interesting as the advent of the social reporting might be in the 1970s, this was followed by a sudden decline in the application and popularity of the phenomenon in the 1980s, and its consequent resurgence in 1990s (Blowfield & Murray, 2011).

In taking this debate forward, Owen and O'Dwyer (2008) have argued that the coming of the 1980s is a precursor to sharp wane in voluntary corporate social and environmental reporting initiatives, ostensibly as a result of declining public interest in the phenomenon as the "New Right" policies of the Thatcher–Reagan period unremittingly held sway. Also, implicated in the decline of social and environmental disclosure in this period were macroeconomic issues globally—particularly in the USA and UK—which generated a cavalcade of market-driven capital market liberalisation movements including policies and acts that wanted to legislate social and environmental reporting out of existence (Blowfield & Murray, 2011). In addition to this, global radicalisation of public management, a metonym for managerialism, "a revolution fuelled by an interest in large-scale governmental reform" (Terry, 1988, p. 195), underpinned liberation management and market-driven management. This movement was aimed at ensuring efficiency, cost saving, and lean government in the management and provision of public goods (Pollitt, 1990).

Nevertheless, the resurgence of social and environmental reporting in the 1990s was essentially pioneered and enabled by activities of organisations such as the CEO-led global alliance called World Business Council for Sustainable Development and the UK's Institute of Social and Ethical Accountability and Council of Economic Priorities as well as related efforts in this regard (Belal, 2008). In the USA, this saw the passing of the Comprehensive Environmental Response Compensation and Liability Act (CERCLA) (otherwise known as superfund). From the academic perspective, writers and scholars began to harp on the internecine relationship between business and society, which culminated in global call for responsiveness, accountability, and transparency in CSR disclosure (and practice) (Carroll, 1991). Of particular note is Frederick's (1994) CSRII typology, which deviates significantly from CSR1. The former is about responsibility, while the latter deepens the rhetoric of responsibility by linking it to responsiveness, which is more of proactive CSR and less of corporate green-washing.

In a similar vein, from enterprise–employee perspective, analogous initiatives for businesses to be more accountable and transparent in their CSR reporting gave rise to the establishment of *socialbilanz* in Germany and *bilan social* in France. *Socialbilanz* was introduced in Germany in 1977 as a move to protect the rights of workers, their benefits and conditions that ensure employers are responsible to them. It is a way of enhancing corporate responsibility and accountability. In France, *bilan social* is produced by companies to indicate their performance and commitment towards environmental wholesomeness. According to Owen and O'Dwyer (2008), of particular note is "... Deutsche Shell whose 'goal directed' reporting employed quantitative indicators to describe the attainment of a wide range of social objectives and attempted to integrate such data with traditional financial accounting" (p. 386–7).

Apart from the foregoing, incidence of oil spills and environmental disaster globally such as Bhopal, Rhine spill, Exxon Valdez, Ogoni (Niger delta) environmental crisis, and a plethora of socio-environmental issues in the world over have all precipitated a rethink of social accounting (Blowfield & Murray, 2011). To this end,

> over the last decade, accounting researchers have turned their attention to the nature of accounting as a social phenomenon...there have been calls for further research into the question of social factors influencing the historical development of contem-porary accounting practices and the social consequences that these have produced (Parker, 1989, p. 343).

Despite the global call for a re-conceptualisation of social disclosure as well as influential journals such as *Accounting, Auditing & Accountability Journal* and *Critical Perspectives on Accounting* among others that were set up to provoke and extend debate on this, it is generally accepted by commentators/scholars (Blowfield & Murray, 2011; Bebbington, 1997; Baker, 2010) that social disclosure is subjected to "managerial capture" (Power, 1991). Apart from the "managerialist slant to the practice of social accounting" (Owen et al., 2000, p. 85), it is also in the shadow of systems theory, which is a way in which dominant and mainstream bourgeois ideologies (in the Marxian parlance) are reified in organisational practice, one of which is "managerial capture" (Power, 1991). The following paragraphs will look at motivations for CSR disclosure.

Motivation for CSR Disclosure

It is worth noting that although CSR disclosure standards and benchmarks are not formulaic and consistent in terms of universal ways in which they are prepared and utilised, companies gain a lot from engaging in CSR disclosure (Deegan, 2002). Thus, companies disclose CSR activities "in order to be accountable to their constituents—their stakeholders, or those agencies that fall within the company's sphere of influence" (Holland & Gibbon, 2001, p. 279; Amaeshi et al., 2006). It is to this end that Nielsen and Thomsen (2007, p. 278) have argued that the motivation to provide an account of a company's activities is determined not only by external factors but also by internal issues including the enhancement of the company's image and the positive effect of this on the bottom line through an improved market share, both of which will have a positive effects on shareholder values (Crane et al., 2008). It is also propelled by a wish to reduce risk, particularly for those companies whose organisational activities are perceived to be environmentally or socially harmful. Given that CSR disclosure is essentially voluntary exercise, corporate self-selection of criteria or indicators may be detrimental to such organisation. Hence, what companies say and how they say it in terms of appropriate discourse ethics and strategies are central to stakeholder conceiving their actions as legitimate or not (Crane et al., 2008; Siltaoja, 2009; Siltaoja & Onkila, 2013).

CSR Disclosure and Organisational Legitimacy

Legitimacy is societal approval of corporate actions (Suchman, 1995). To this end, corporations try to achieve legitimacy at all cost, which sometimes brings negative repercussions. In a bid to achieve legitimacy through CSR disclosure, companies have run the risk of rather achieving contrary effects of their actions (Morsing & Schultz, 2006). This is what Ashford and Gibbs (1990) tagged "self-promoter's paradox" (p. 188). This concept states that firms over-emphasise their CSR activities and commitment in a way that oftentimes leads to contradictory expectations. Accordingly, as Christensen and Cheney (2000) contend, corporate managers become "self-seduced and self-absorbed" in their effort to disclose their CSR issues, which amply manifests in the way and manner they do this—by using what Habermas (1984) calls instrumental (strategic) language and discourse rather than communicative language (normative or post-positivistic discourse). The former is exclusive of stakeholder engagement in its operationalisation; while the latter is dialogic and accountable. Despite having to decide on what to disclose, firms have other ways or alternative formats to report CSR issues (Maguire, 2011). Apart from CSR reports, companies are increasingly choosing to integrate their sustainability reports, environmental reports, and similar disclosures in their annual reports (Nielsen & Thomsen, 2007) for accountability, legitimacy, and transparency. Also,

CSR disclosure impacts on organisational performance and stakeholder engagement, which shall be considered in turn.

CSR Disclosure and Organisational Performance

The work of Ullmann (1985) is usually used when interrogating the relationship between CSR disclosure and firm's financial performance, stakeholder power and strategic position (posture) of firms. The first element of this theory—stakeholder power—stipulates that firms will give greater salience (Mitchell et al., 1997) to issues that relate to stakeholders with more powers than others. Stakeholder salience is the degree of importance that firms attach to stakeholders' concerns. The second integer is performance, which is associated with amount of disclosure firms make as a consequence of relative financial positions of firms. So, firms in a good financial state would make more disclosure as opposed to others that are not as financially viable as them (Ullmann, 1979, 1985). The final element—strategic posture—considers disclosure as a function of relative strategic position of firms. Thus, a firm with active strategic position can influence stakeholders by making more disclosures for strategic gains. This process has been tested empirically by Roberts (1992) in which case he empirically observed that there is a positive correlation between socially performing organisation and high level of CSR disclosure.

CSR Disclosure and Stakeholders

Although the stakeholder theory had been conceived earlier, it was Freeman (1984) that popularised the model as a useful heuristics in apprehending "contributors" to organisational wealth. Given the imperative of this line of argument, the discipline of management—particularly social and environmental accounting—is recently awash with a growing trend in studies that are aimed at understanding how the entire gamut of corporate accountability "… is controlled by management and as such the legitimacy of the scope of the reporting …" (Sweeney & Coughlan, 2008, p. 114). Interestingly, CSR disclosure is one effective means of apprehending how social and environmental matters are disclosed to other stakeholder groups. Extant literature on stakeholder accountability and engagement (Morsing & Schultz, 2006; Bebbington, 1997) suggests that each stakeholder group has a "stake" in the way organisations' wealth is shared (Sweeney & Coughlan, 2008). Therefore, it is imperative that organisations look beyond their narrow confines of economic benefit and take into cognisance other stakeholders' interests, apart from protecting only interests of shareholders (stockholders) (Clarkson, 1995; Freeman, 1984).

Further to understanding the stakeholder concept, some theories have been developed around Freeman's (1984) initial theory. These theories include Clarkson's (1995) risk-based analysis of stakeholder theory, Donaldson and Preston's (1995) stakeholder theory of corporations, and Mitchell, Agle, and Wood's (1997) stakeholder salience model. Thus, analysing how CSR disclosure gives priority as well as levels of importance to various stakeholders is central to understanding managerial capture. In the case of Nigeria, the community people in the Niger delta region, whose means of livelihood is in jeopardy given the presence of oil and gas companies have perennially criticised the manner in which these companies report CSR. Nigerian stakeholders maintain that CSR disclosure rather than being instruments of engagement are purely instrumental (managerial) by design (Ogiri et al., 2012).

CSR Disclosure and Stakeholders: Understanding Who Matters to the Firm

Freeman (1984) made a Kantian recommendation—categorical imperative assertion—which states that each "… stakeholder has a right to be treated as an end, not a means" (Wijnberg, 2000, p. 331). In taking this debate further, Freeman and Evan (1990) proposed a re-articulation of the philosophy, which urges firm to take cognisance of co-ordinating stakeholders' interests and aspirations. This mirrors Freeman's (Freeman, 1984) "doctrine of fair contracts". For Wijnberg (2000), this is tantamount to balancing the interests of stakeholders in CSR strategy through the instrumentality of legitimacy rather than power and urgency. The former is responsible for why greater salience is given to primary stakeholders and shareholders rather than external or secondary stakeholders (Clarkson, 1995; Freeman, 1984; Mitchell et al., 1997).

The instrumental perspective to stakeholder theory finds incarnation in stifling stakeholders' voice in CSR reporting processes (Morsing & Schultz, 2006). It also gives priority to internal, powerful, and influential stakeholders (Mitchell et al., 1997). Unlike the normative perspective, which prescribes who ought to be considered as a stakeholder including ethics of right or wrong in relation to stakeholding, instrumental stakeholder theory "… suggests that firms are interested in stakeholders because of perceived benefit in terms of improved financial performance" (Belal, 2008, p. 19). In what can be considered a bedlam of effort to increase shareholder gains and to further firms' interests, instrumental lens to stakeholder theory diminishes legitimacy and accountability in CSR reporting. It panders to the allure of protecting shareholders' interest at the expense of other stakeholders, whose stakes are part of the engine of the firm (Freeman, 1984). Relative power and importance of stakeholders are major determinants of how firms treat them in relation to CSR reporting and practice. From a managerial standpoint, firms report CSR issues based on relative power, influence, and attributes of various stakeholders. It is on the strength of this that the present study proposes normative approach to corporate stakeholding, which is considered in detail in the following paragraphs.

Prospecting for Normative (Post-positivistic) Stakeholder Theory for Legitimacy

A normative conceptualisation aims to discover what should be and would include perspectives like "organisations should follow corporate governance principles/ standards" or "managers ought to act in a specific way to avoid conflicts of interests". Firms that believe in normative practices play a major role in society by raising public consciousness, addressing socio-economic imbalance and environmental causes, as well as bringing social change, and helping vulnerable people. Freeman (1984) conceptualised a normative essential for stakeholder theory premised on the notion that values creation process and activity is based on a contractual engagement involving customers, financiers, communities, employees, and suppliers. The normative essential or what he referred to as "normative core" in understanding stakeholder theory is premised on value creation activity with respect to organisations' survival and meeting their strategic objectives, which are contractual process among customers, financiers, suppliers, employees, and communities. This normative core rotates along the axis of the Rawlsian liberal idea of autonomy, justice, and solidarity (Friedman & Miles, 2002; Rawls, 1971). In this direction, Freeman's overriding intention was to offer some clarity on Goodpaster's (1991) "separation thesis", which entails the dialectics of business devoid of ethics or ethics devoid of business. This process of conceptualising business in contradictory philosophical moorings deals with foregrounding morality and legitimacy in business in relation to managerial preferences, stakeholder power, and salience (Mitchell et al., 1997).

In Freeman's (1984) thinking, firms implicitly enter an agreement or contract with other stakeholders, whose interests ought to be considered for organisational performance and legitimacy (Suchman, 1995). Beyond the mantra of institutional legitimacy (Scott, 2003; Meyer & Scott, 1983) and strategic legitimacy (Dowling & Pfeffer, 1975) the "licence to operate" dictum is central to what Mele and Schepers (2013) consider as "networked legitimacy", which chimes with Donaldson & Preston's (1995) normative stakeholder theory. Therefore, as Kaler (2003) notes the normative approach chimes with the use of stakeholder theory in stating what the "function" of corporations should be and the moral or ethical parameters "they should follow with regard to their operation and management" (p. 73), which are obviously articulated in the social contract.

The contract technically defines what firms should do in relation to the role of business *in* society as opposed to business *and* society (Siltaoja & Onkila, 2013). As a consequence, the contract spells out companies' rules of engagement, which are moderated by their *modus operandi* on what is to be decided and done behind the Rawlsian "veil of ignorance" (Rawls, 1971). To this end, stakeholder representatives presumably are self-interested entities who potentially have a good knowledge of corporations' current and future initiatives and plans. They are also expected to understand the consequences of disparate organisations' plans in relation to success and failure, including having a fair knowledge of other institutional frameworks that exist (Friedman & Miles, 2005). At the opposite end is strategic or instrumental/

managerial stakeholder theory; its relationship to how stakeholders are considered in organisations in terms of who has what stake or interest representation for legitimacy will be explained in the following paragraphs.

Conclusion

This chapter has explored CSR disclosure, motivation for disclosure, and who matters to the firm and subsequent argument for normative organisational practice as a consequence. Specifically, the paper has outlined the meaning of CSR, the transition from CSR to CSR reporting and rationale for disclosure as well as the rise of CSR disclosure investigating the trajectory from the traditional reporting/accounting to the present era, an epoch of social accounting and environmental reporting for sustainable, normative organisational practice. Additionally, the motivation for CSR disclosure has been given attention here including CSR disclosure and stakeholders by understanding who matters to the firm. Finally, there is prospect for normative/post-positivistic stakeholder theory for legitimate, sustainable, and normative organisational behaviour and practice.

References

Alazzani, A., Aljanadi, Y., & Shreim, O. (2019). The impact of existence of royal family directors on corporate social responsibility reporting: A servant leadership perspective. *Social Responsibility Journal, 15*(1), 120–136.

Amaeshi, K., Adi, B., Ogbechie, C., & Amao, O. (2006). Corporate social responsibility in Nigeria: Western mimicry or indigenous practices? *ICCR Research Paper Series, 39*, 1–39.

Amnesty. (2015). *Niger Delta: Shell's manifestly false claims about oil pollution exposed, again*. Amnesty International.

Asechimie, D. P. S. (1992). *A history of Accounting in Nigeria*. CSS Press.

Ashford, B. E., & Gibbs, B. W. (1990). The double-edge of organisational legitimation. *Organisation Science, 1*(2), 177–194.

Baker, M. (2010). Re-conceiving managerial capture. *Accounting, Auditing & Accountability Journal, 23*(7), 847–867.

Bebbington, J. (1997). Engagement, education and sustainability: A review essay on environmental accounting. *Accounting, Auditing and Accountability Journal, 10*(3), 365–381.

Belal, A. R. (2004). *The prospects for corporate social reporting (CSR) in Bangladesh*. Doctoral thesis. Sheffield: University of Sheffield.

Belal, A. R. (2008). *Corporate social responsibility reporting in developing countries: The case of Bangladesh*. Routledge.

Blowfield, M., & Murray, A. (2011). *Corporate responsibility*. Oxford University Press.

Carroll, A. B. (1991). The pyramid of corporate social responsibility: Toward the moral management of organisational stakeholders. *Business Horizons*, July–August, 39–48.

Carroll, A. B., & Shabana, K. M. (2010). The business case for corporate social responsibility: A review of concepts, research and practice. *International Journal of Management Review, 12*(1), 85–105.

Chantziaras, A., Dedoulis, E., Grougiou, V., & Leventis, S. (2020). The impact of religiosity and corruption on CSR reporting: The case of US banks. *Journal of Business Research, 109*, 362–374.
Christensen, L. T., & Cheney, G. (2000). *Self-absorption and self-seduction in the Corporate Identity Game* (pp. 246–270). Oxford University Press.
Clarkson, M. B. E. (1995). A Stakeholder framework for analysing and evaluating corporate social performance. *Academy of Management Review, 20*(1), 92–117.
Crane, A., & Matten, D. (2010). *Business ethics: Managing citizenship and sustainability in the age of globalization*. Oxford.
Crane, A., McWilliams, A., Matten, D., Moon, J., & Siegel, D. S. (Eds.). (2008). *The Oxford handbook of corporate social responsibility*. Oxford University Press.
Crowther, D., & Aras, G. (2008). Governance and sustainability: An investigation into the relationship between corporate governance and corporate sustainability. *Management Decision, 46*(3), 433–448.
Deegan, C. (2002). The legitimising effect of social and environmental disclosures – A theoretical foundation. *Accounting, Auditing and Accountability Journal, 15*(3), 282–311.
Disu, A., & Gray, B. (1998). An exploration of social reporting and MNCs in Nigeria. *Social and Environmental Accounting Journal, 18*(2), 13–15.
Donaldson, T., & Preston, L. E. (1995). The stakeholder theory of the corporation: Concepts, evidence and implication. *Academy of Management Review, 20*(1), 65–91.
Dowling, J. B., & Pfeffer, J. (1975). Organisational legitimacy: Social values and organizational behaviour. *Pacific Sociological Review, 18*(1), 122–136.
Egbon, O., & Mgbame, C. O. (2020). Examining the accounts of oil spills crises in Nigeria through sensegiving and defensive behaviours. *Accounting, Auditing & Accountability Journal, 33*(8), 2053–2076.
Elkington, J. (1997). *Cannibals with forks: The ripple bottom line of the 21st century*. Capstone Publishing.
Elliott, B., & Elliott, J. (2006). *Financial accounting, reporting and analysis* (2nd ed.). Pearson.
Emeseh, E., & Songi, O. (2014). CSR, human rights abuses and sustainability report accountability. *International Journal of Law and Management, 56*(2), 136–151.
Eweje, G. (2007). Multinational oil companies' CSR initiatives in Nigeria. *Managerial Law, 49*(5/6), 218–235.
Ferguson, J., Sales de Aguiar, T. R., & Fearfull, A. (2016). Corporate response to climate change: Language, power and symbolic construction. *Accounting, Auditing & Accountability Journal, 29*(2), 278–304.
Frederick, W. C. (1994). From CSR1 to CSR2: The maturing of business and society thought. *Business and Society, 33*(2), 150–164.
Freeman, R. E. (1984). *Strategic management: A stakeholder approach*. Pitman.
Freeman, R. E., & Evan, W. M. (1990). Corporate governance: A stakeholder interpretation. *Journal of Behaviour Economics, 19*, 337–359.
Friedman, A. L., & Miles, S. (2002). Developing stakeholder theory. *Journal of management studies, 39*(1), 1–21.
Friedman, A. L., & Miles, S. (2005). Developing stakeholder theory. *Journal of Management Studies, 39*(1), 1–21.
Frynas, J. G. (2005). The false promise of corporate social responsibility: Evidence form multinational oil companies. *International Affairs, 81*(3), 581–598.
Goodpaster, K. E. (1991). Business ethics and stakeholder analysis. *Business Ethics Quarterly, 1*(1), 53–73.
Gray, R. (2002). The social accounting project and accounting, organisations and society: Privileging engagement, imaginings, new accountings and pragmatism over critique? *Accounting, Organisations and Society, 27*(7), 687–708.

Gray, R., Kouhy, R., & Lavers, S. (1995). Corporate social and environmental reporting: a review of the literature and a longitudinal study of UK disclosure. *Accounting, auditing & accountability journal, 8*(2), 47–77.

Gray, R., Owen, D., & Adams, C. (1996). *Accounting and accountability: Changes and challenges incorporate social and environmental reporting*. Prentice Hall.

Guthrie, J., & Parker, L. D. (1989). Corporate social reporting: A rebuttal of legitimacy theory. *Accounting and Business Research, 19*(76), 343–352.

Habermas, J. (1984). *The theory of communicative action: Reason and the rationalisation of society*. Thomas McCarthy (Trans.). Massachusetts: Beacon Press.

Higgins, C., & Walker, R. (2012). Ethos, logos, pathos: Strategies of persuasion in social/environmental reports. *Accounting Forum, 36*(3), 194–208.

Holland, L., & Gibbon, J. (2001). Processes in social and ethical accountability: External reporting mechanisms. In J. Andriof & M. McIntosh (Eds.), *Perspectives on corporate citizenship* (pp. 278–295). Sheffield.

Hopper, T., Tsamenyi, M., Uddin, S., & Wickramasinghe, D. (2009). Management accounting in less developed countries: What is known and needs knowing. *Accounting, Auditing & Accountability Journal, 22*(3), 469–514.

Idemudia, U. (2010). Corporate social responsibility and the rentier Nigerian state: Rethinking the role of government and the possibility of corporate social development in the Niger Delta. *Canadian Journal of Development Studies, 30*(1–2), 131–153.

Idemudia, I., & Ite, U. E. (2006). Corporate-community relations in Nigeria's oil industry: Challenges and imperatives. *Corporate Social Responsibility and Environmental Journal, 13*(4), 194–206.

Ite, U. E. (2004). Multinationals and corporate social responsibility in developing countries: A case study of Nigeria. *Corporate Social Responsibility and Environmental Journal, 11*(1), 1–11.

Kaler, J. (2003). Differentiating stakeholder theories. *Journal of Business Ethics, 46*(1), 71–83.

Kamla, R. (2007). Critically appreciating social accounting and reporting in the Arab middle east: A postcolonial perspective. *Advances in International Accounting, 20*, 105–177.

Karl, T. L. (1997). *The paradox of plenty: Oil booms and petro-states*. University of California Press.

KPMG. (2017). *KPMG survey of corporate social responsibility reporting*. http://www.home.kpmg›Home›Insights. (Accessed June 2023).

Maguire, M. (2011). The future of corporate social responsibility reporting. *Issues in Brief.* https://www.bu.edu/pardee/publications-library/issues-in-brief-no-19/. (Accessed 20/07/2023).

Mele, V., & Schepers, D. (2013). E pluribus unum? Legitimacy issues and multi-stakeholder codes of conduct. *Journal of Business Ethics, 118*(3), 561–576.

Meyer, J. W., & Scott, W. R. (1983). *Organisational environments: Ritual and rationality*. Sage.

Mitchell, R. K., Agle, B. R., & Wood, D. J. (1997). Towards a theory of stakeholder identification and salience. *Academy of Management Review, 22*(4), 853–886.

Morsing, M., & Schultz, M. (2006). Corporate social responsibility communication: Stakeholder information, response and involvement strategies. *Business Ethics: A European Review, 15*(4), 323–338.

Nielsen, A. E., & Thomsen, C. (2007). Reporting CSR – What and how to say it? Corporate communication. *An International Journal, 12*(1), 25–40.

Nwagbara, U., & Alhasan, Y. (2021). *Microfinance and sustainable development in Africa*. IGI Global Publishing.

Nwagbara, U., & Belal, A. (2019). Persuasive language of responsible organisation? A critical discourse analysis of corporate social responsibility (CSR) reports of Nigerian oil companies. *Accounting, Auditing & Accountability Journal, 32*(8), 2395–2420.

Ogiri, I. H., Samy, M., & Bampton, B. (2012). Motivations of legitimacy theory for CSR reporting in the Niger Delta region of Nigeria: A theoretical framework. *African Journal of Economic and Sustainable Development, 1*(3), 265–283.

Okoye, A. (2009). Theorising corporate social responsibility as an essentially contested Concept: Is a definition necessary? *Journal of Business Ethics, 89*(4), 613–627.

Owen, D. (2005). CSR after Enron: A role for the academic accounting profession? *European Accounting Review, 14*(2), 395–404.
Owen, D. L., & O'Dwyer, B. (2008). Corporate social responsibility: The reporting and assurance dimension. In Andrew Crane et al. (Ed.), *The Oxford handbook of corporate social responsibility*, pp. 384–410.
Owen, D., Swift, T., Humphrey, C., & Bowerman, M. (2000). The new social audits: Accountability, managerial capture or the agenda of social champions? *European Accounting Review, 9*(1), 81–98.
Parker, R. H. (1989). Importing and exporting accounting: The British experience. In A. G. Hopwood (Ed.), *International Pressures for Accounting Change*. Prentice Hall.
Pollitt, C. (1990). *Managerialism and the public services*. Blackwell.
Power, M. (1991). Auditing and environmental expertise: Between protest and professionalism. *Accounting, Auditing and Accountability Journal, 4*(3), 30–42.
Rawls, J. (1971). *A theory of justice*. Harvard University Press.
Roberts, R. W. (1992). Determinants of corporate social responsibility disclosure: An application of stakeholder theory. *Accounting, Organisation & Society, 17*(6), 595–612.
Ross, M. L. (1999). The political economy of the resource curse. *World Politics, 51*(2), 297–322.
Scherer, A. G., & Palazzo, G. (2011). The new political role of business in a globalised world: A review of a new perspective on CSR and its implications for the firm, governance, and democracy. *Journal of Management Studies, 48*(4), 899–931.
Scott, W. R. (2003). *Organisations: Rational, natural and open systems* (5th ed.). Prentice-Hall.
Siltaoja, M. E. (2009). On the discursive construction of a socially responsible organisation. *Scandinavian Journal of Management, 25*, 191–202.
Siltaoja, M. E., & Onkila, T. J. (2013). Business in society or business and society: The construction of business–society relations in responsibility reports from a critical discursive perspective. *Business Ethics: A European Review, 22*(4), 357–373.
Suchman, M. C. (1995). Managing legitimacy: Strategic and institutional approaches. *Academy of Management Review, 20*(3), 571–610.
Sweeney, L., & Coughlan, J. (2008). Do different industries report Corporate Social Responsibility differently? An investigation through the lens of stakeholder theory. *Journal of Marketing Communications, 14*(2), 113–124.
Terry, L. D. (1988). Administrative leadership, neo-managerialism, and the public management movement. *Public Administration Review, 58*(3), 194–200.
Ullmann, A. A. (1979). Corporate social reporting: Political interests and conflicts in Germany. *Accounting, Organisations and Society, 4*(1/2), 123–133.
Ullmann, A. A. (1985). Data in search of a theory: A critical examination of the relationships among social performance, social Disclosure, and economic performance of U.S. firms. *Academy of Management Review, 10*(3), 540–577.
UNEP. (2011). *Environmental assessment of Ogoniland, United Nations Environment Programme (UNEP)*. UNEP.
Visser, W. (2013). *The age of responsibility: CSR 2.0 and the New DNA of Business*. Wiley.
Wijnberg, N. M. (2000). Normative stakeholder theory and Aristotle: The link between ethics and politics. *Journal of Business Ethics, 25*(4), 329–342.
Zwetsloot, G. I. J. M. (2003). From management systems to corporate social responsibility. *Journal of Business Ethics, 44*, 201–208.

Uzoechi Nwagbara is a professor of management at ICT University. He has BA and MA in English as well as MSc and PhD in Human Resource Management and Management, respectively, from the University of Wales, and MEd in Higher Education as well as PG Cert in Academic Practice from the University of Sunderland, UK. He is a Fellow of the Higher Education Academy (FHEA); Member Chartered Management Institute (MCMI); Member Institute of Leadership & Management (MInstLM); and Member Community Development Exchange (MCDE); and also affiliated with British Academy of Management (BAM), European Association for the Study of Literature, Culture, and the Environment (EASLCE), Postcolonial Studies Association (PSA), and Royal African Society (RAS). Dr Nwagbara has over 35 years of teaching experience and has taught in higher education for upwards of 25 years. He is external examiner for University of Cumbria and University of Bolton. He has supervised over a dozen doctoral students to completion and published well over 100 publications in high impact journals in management, leadership, CSR, sustainability, employment relations, and human resource management. Dr Nwagbara has examined PhD students internally and externally in various universities including University of Bedfordshire, Cardiff Metropolitan University, and University of the West of Scotland amongst others. He holds Visiting Teaching Positions at the University of Sunderland in London, Coventry University London, London School of Commerce, University of the West of Scotland, The University of Law UK, University of Worcester, University of Wales Trinity Saint David, and University of Plymouth as well as Visiting Professorship (management), Coal City University. He is on the editorial boards of many peer-reviewed, international journals including *Economic Insights—Trends & Challenges*, *Journal of Pan African Studies* and *Journal of Sustainable Development in Africa*.

Dr Yahaya Alhassan is a Principal Lecturer in Finance and Strategic Management at the University of Sunderland in London. He is also the Head of Academic Operations responsible for quality assurance and new programme development. He holds PhD in Business Management from Anglia Ruskin University, MBA in Finance as well as MA in Education from the University of Sunderland. He also holds BA in Business Management and a Senior Fellow of the Higher Education Academy (HEA). He has over 20 years' teaching and management experience developed during roles in the higher education sector and the consultancy industry. Dr Alhassan is a researcher, author, finance and strategic management lecturer, and consultant and has published over seventeen publications including books, book reviews, journal articles, and book chapters. Some of Dr Alhassan's publications have appeared in the UK ABS 3*** rated Journals. Dr Alhassan regularly presents papers at conferences and has presented over ten papers at various conferences since 2019. His main research interests are around development finance, microfinance, entrepreneurial finance, ethnic minority entrepreneurship and microbusiness development. His current research interests focus on the nexus between financial technology and sustainable entrepreneurship, investment in artificial intelligence, and firm growth. He has also provided consultancy services to organizations nationally and internationally that positively influenced policies, practices, and reached audiences beyond academia.

Ngozi Ibeawuchi, BA, MA is a doctoral candidate at the University of Greenwich. She is currently lecturing at Coventry University London and Regent College, UK. Her areas of expertise include leadership, special education, and CSR and she has published in several journals. Her doctoral focus is on special education and the role played by transformational leadership in the process of making educational leadership more inclusive.

Jacyntha Stewart, MA, is Social Impact & Investment Consultant and Founder & CEO For Business Sake, United Kingdom. https://orcid.org/. e-mail: info@forbusinesssake.org. She is currently pursuing her doctorate at University of West London, UK, specializing in institutional racism in the workplace.

Chapter 4
The Role of Transformational Leadership in Promoting Corporate Governance Effectiveness in Nigeria

Chigozie A. Ugwoji

Introduction

With the growing importance of ethical business practices and the need for accountable and transparent leadership, the role of transformational leaders in driving corporate governance practices becomes paramount. This study examines how transformational leadership could engender accountable modus operandi of corporate governance globally and in Nigeria, which has a high incidence of corporate governance maleficence (Adegbite, 2012; Okike, 2007; Klapper & Love, 2002). Corporate governance is assumed as an essential requirement for social, economic, and political developments (Gisselquist, 2012) of any given state. However, corporate failures, occur when firms are unable to meet up with required standards or when they give rise to negative impacts borne by leaders. For instance, the BP oil spillage disaster in the Gulf of Mexico in 2010 and collapse of global financial system in 2007 and 2008 highlight the profound and far-reaching impacts of poor leadership activities. The argument of this study is that transformational leadership has the potential to create a more reliable, accountable, and best-practice conduct within an organisation, which in turn translates to business successes. Transformational leaders therefore have strong visions for the success of organisations and effectively communicate this vision to their team members. By doing so, they inspire a sense of purpose and encourage workers, to align their goals and actions with the organisation's objectives/purpose. Additionally, they demonstrate integrity and strong work ethics, which in turn encourages workers to follow suit. This ethical conduct promotes transparency, accountability, and responsibility in decision-making

C. A. Ugwoji (✉)
University of Sunderland in London, London, UK

Brunel University, London, UK
e-mail: chi.ugwoji@sunderland.ac.uk

U. Nwagbara et al. (eds.), *Corporate Social Responsibility Disclosure in Developing and Emerging Economies*, CSR, Sustainability, Ethics & Governance, https://doi.org/10.1007/978-3-031-61976-2_4

throughout the organisation, hence promoting business goals and a sustainable business environment. More so, transformational leaders help in fostering a culture of loyalty, commitment, and high performance, which could serve as a springboard for a strengthened, accountable corporate governance practice, and loss of confidence in corporate businesses, especially in Nigeria (Okike, 2007).

Arguably, attention has been paid to issues and loss of confidence in corporate governance over the years (Adegbite, 2012; Cadbury, 2002), with considerable interest both in the business world and in scholarly discourses (Nwagbara & Ugwoji, 2015; Claessens & Yurtoglu, 2013) as well as a growing phenomenon. This is informed by the impacts of corporate maleficence and failures globally. Such incidents include Enron, WorldCom, Parmalat scandal. Northern Rock and Lehman Brothers issues, amongst others, have thrown accountability and corporate governance into turmoil. Since the Cadbury Report of 1992, extensive attention has been made globally to interrogate why companies fail, which financial analysts traced to misconduct and corporate malfeasance, triggered this scholarship pattern in terms of corporate governance. In Nigeria, however, incessant bank failures, privatisation, mergers and acquisitions, financial impropriety, and bad leadership practices have occasioned accountability and credibility gaps in corporate governance culture by corporate businesses (Nwagbara & Ugwoji, 2015; Nwagbara, 2014; Emeseh & Songi, 2014). In response to intense accountability issues and corporate failures, businesses are required to implement practices effectively to meet targets and exceed performances (Mammassis & Kostopoulos, 2019), which can only be achieved with influence of transformational leadership style/ behaviour in managing and sustaining corporate business (Sadeghi & Pihie, 2012).

Corporate Governance and Corporate Behaviour in Nigeria

Corporate governance in Nigeria can be traced to the colonial days of the British administration that imposed Anglo-Saxon base system of corporate law and regulation before Nigerian independence (Adegbite et al., 2013). This means that the conduct and governance of Nigeria's firms were drawn within the provision and legislation of the West (Britain), leaving Nigeria with no other option than to inherit Anglo-Saxon framework of corporate governance (Okike, 2007). After independence, the Nigerian government replaced the Companies Ordinance of 1922 with the 1968 Companies Act that modelled the UK Companies Act of 1948. This implies that all the reforms in law and legal system are fashioned towards the Anglo-Saxon model. Following the international standards and codes of best practices of corporate businesses globally, the World Bank (WB) and International Monetary Fund (IMF), called for internationally recognised standards and codes as well as Reports on the Observance of Standards and Codes (ROSC) for more responsible business conduct (Okike, 2007).

In this line of thought, a motion was set to prepare reports on the observance of standards and codes (Accounting and Auditing) in Nigeria; following the poor managerial practices in-listed companies which were manned by the World Bank team between November 2003 and March 2004 (Vanguard, 2007). Prior to ROSC developments, the urgent need to address corporate governance was recognised in September 1998 (being the 28th Annual Accountants Conference organised by the Institute of Chartered Accountants of Nigeria (ICAN)), which had 'Corporate Governance' as the main thrust, reiterating the commitment of the profession including 'trust, transparency, accountability and honesty in the management of the nation's resources', and to 'disprove and enlighten publics that corporate failures are not synonymous with audit failures' (Okike, 2007). However, weak corporate regulation, passive attitude of the political leadership, multiplicity of corporate governance codes, and corporate corruption (Nakpodia, 2016; Adegbite, 2012) have traditionally been at the centre of corporate businesses and governance issues in Nigeria. This is why Okike (1994 & 2007) suggests that the presumed corruption within the accounting and finance profession in Nigeria led to the promulgation of the 'curious' auditing regulation (such that auditors' reports be countersigned by legal practitioners). Ironically, the process of introducing the 'curious' requirement into law was itself perceived as corrupt. Research such as Nakpodia (2016) affirms that lack of morality in Nigeria is largely associated with the high growing incidences of poverty, incompetent and corrupt leadership, lack of respect for human rights and rule of law. This crisis in governance implicates (a) past military governments, (b) inexperience in public service, (c) corrupt and bad leadership and governance, and (d) weak and corrupt political processes (Adegbite et al., 2013).

Arguably, Nigeria has a history of a significant number of high-profile and often unthinkable malpractices/frauds that have been perpetrated by leaders/managers and directors of both listed and non-listed/private and public corporations and individuals alike (Adegbite, 2012). For example, in the early 1990s, the country's financial sector experienced a major instability which resulted in the collapse of several financial institutions (ROSC, 2004). Also, a recent survey conducted found that corporate governance lapses were significantly responsible for the collapse of over 70% of defunct companies in Nigeria over the last two decades (Adegbite, 2015). For instance, the Nigeria Deposit Insurance Corporation (NDIC) highlighted banks failures to include economic depression interference of board members (32.1%), (25%), Bad credit policy (25%), and political crisis (17.9%) (Adegbite, 2015), as well as dissolution of Nigeria Telecommunication Limited (NITEL) and Nigeria Mobile telecommunication Limited (MTEL) in 2014 after attempts to privatise the companies proved unsuccessful (Adegbite, 2015).

These failures were attributed to weak managerial issues, lack of sustainable managerial policies, management/leadership malpractices, and harsh operating conditions, resulting in many business failures (Kazemian et al., 2017), as well as poor governance system amongst the ruling class. The systemic corporate failures in Nigeria undermined customers' trust and highlighted the importance of good corporate governance for corporate vitality and economic stability. There are corporate governance legislation promulgated by the government to revitalise business practices and enhance accountability in businesses, however, the mechanisms for enforcement and compliance are infective and has been historically feeble and poorly managed by powers that be (Adegbite et al., 2013; Adegbite, 2012, 2015; Okike, 2007; Hirschman, 1984). This is as a consequence of bribery and corruption which has taken centre stage in the Nigerian business environment. For example, the issue of non-compliance was evident in the accounting and auditing practices in Nigeria that suffered from institutional weaknesses in regulation, compliance, and enforcement of standards and rules (ROSC, 2004, p. ii). Similarly, Codes of Conduct and Best Practices in corporate governance were enacted to arrest issues of poor corporate governance (Akinpelu, 2012) and promote fairness to shareholders, clear accountability, transparency in financial and non-financial reporting, and compliance with the law were not enforced (Coombes & Chiu-Yin Wong, 2004).

Notwithstanding the presence of codes and best practices to checkmate corporate fraud and malpractices in Nigeria, yet good corporate governance behaviour/practices are still lacking in corporate business. This is as a result of endemic corruption rooted in Nigeria's corporate businesses as well as infective and weak institutional, legal, and regulatory frameworks (Wilson, 2006). This means that non-identification of corruption as an issue is an obvious omission in the code of Best Practice. Corruption is so pervasive and embedded in Nigeria's culture, which has turned some public services into somewhat criminal enterprises. Given that the dominant understanding of corporate governance rests on the board (management/leaders), as stated in the Turnbull Report by the Financial Reporting Council of 2014 as cited in Spiers (2017, p. 117); 'the board has ultimate responsibility for risk management and internal control, … willing to achieve its strategic objectives… ensure appropriate culture has been embedded throughout the organization'. This means that leadership provides the motivation and desire to make corporate governance effective as it creates a culture (behaviours needed to embed corporate governance) that uses effective leadership behaviour to achieve commercial competitiveness, business viability, business accountability, and transformed society.

Inept leadership and poor corporate governance: The challenge of leadership—Leadership is a critical element that shapes the success and sustainability of businesses and good societal functioning of the state (Bass, 1999; Bennis, 1995; Kouzes & Posner, 2007; Sergiovanni, 1984; Useem, 2001). Leadership involves inspiring and guiding individuals and teams towards fostering ethical behaviour, decision-making processes, and shared goals for the team and businesses, as well as support to accomplish tasks/goals in an organisation (Eva et al., 2018). This means that leadership is about an interaction with others/followers, a way to enhance personal growth, professional and organisational advancement towards realisation of business objectives and goals; influencing organisation's strategy and corporate culture that leads to business successes and growth (Eva et al., 2018).

Arguably, corporate governance, incorporates practices, system of rules, and processes that ensure fairness, transparency accountability, in organisational operations. Therefore, corporate governance is how businesses can be effectively led that ensures accountability and transparency. This means that leadership creates corporate governance culture of integrity, responsible decision-making that contributes to long-term sustainability and stakeholder trust (Wang & Zhou, 2016), which has a direct correlation to sustainability of business successes (Gao & Zhang, 2019).

Additionally, corporate businesses are established through the formation of effective leadership/Board of Directors roles for critical operational processes of the business, as they communicate vision, direction, and purpose needed for the success of the business (Similarly, Essen et al., 2013; Yukl, 2006). Therefore, a leader propels, enables, and motivates followers and group members towards actualisation of a collective vision to transform businesses and society. However, research has highlighted other possible influences and variables to business performances of the business including entrepreneurial orientation, information technology, and business strategy. However, these variables cannot influence business performance on their own without a leader at the front burner, thus the styles/behaviour of leadership have a significant contribution to business overall performance (Bass & Avolio, 2000). With the recent development of leadership models from traditional to modern—"transformational leadership" has become popular amongst other leadership behaviours that uplift corporate business (Asif et al., 2019). Transformational leaders transform followers' desires and preferences into shared and common interests by sharing values and collective decision-making. Transformational leaders enable followers to work independently for a collective purpose, which can impact accountability, transparency, and transform businesses and society (Hoffmeister et al., 2014).

Transformational Leadership for Better Corporate Governance

Scholars advocate that transformational leadership has highlighted inspiring innovativeness as a basic leadership function (Bass, 1985) that focuses on coaching and transforming others to achieve a higher level of performance (Bass, 1985). This means that transformational leaders focus on developing the potential of their team members. They invest time and effort in coaching, mentoring, and providing constructive feedback. This not only helps members to improve their skills and knowledge but also contributes to the development of a competent and capable workforce. Through continued learning and growth, employees become more equipped to handle the complexities and challenges of corporate governance issues, which elevate their level of consciousness around goals, knowledge-sharing, and achieving more sustainable business performances (Gerard et al., 2017).

Additionally, transformational leaders empower members to become leaders, as moral agents by improving their performance and developing them to their full potential (Northouse, 2021, Bennis, 2010; Tafvelin et al, 2012; Bass & Riggio, 2006; Burns, 1978). This aptitude finds expression in the Leader–Member Exchange (LMX) theory that identifies leader–follower mutual and social exchange relationship as well as follower unique responsibilities and attributes (Yang, 2015; Graen & Uhl-Bien, 1995). For example, transformational leaders assure organisational success by passing on the organisation's goals, mission, and vision to their members (Burch & Guarana, 2014). Research argue that transformational leadership behaviour is suitable for the implementation of business strategy and company performance. Other studies found the heavy influence of transformational leadership behaviour with good corporate performance as measured by financial performance metrics, e.g., stock price, market share, debt-to-equity ratio, return on assets, and earnings per share (Avolio, 2007).

Bass (1985) articulated four dimensions (as the hallmark of Leadership and Performance beyond Expectation) of transformational leadership styles, including Idealised influence, inspirational motivation, intellectual stimulation, and individualised consideration (4Is) (Clarke, 2007). These components have been shown to positively influence various employee outcomes that are vital for organisational functioning and transformation, including job performance, job satisfaction, staff engagement, and staff motivation (e.g., Fisher, 2009; Lawler, 2007; Judge & Piccolo, 2004). The blueprint for effectiveness in this model is that a transformational leader recognises his/her employees' behaviour is not purely driven by standard rewards and punishment but truly motivated by stimulating opportunities including promoting working together towards a common good and purpose.

Intellectual Stimulation

Bass (1985) has argued that transformational leaders stimulate employees' efforts to be committed, innovative and creative, and question the status quo, re-examine traditional ways of thinking and doing things, and suggest new ways to approach work-related problems and challenging their own self-beliefs (Rafferty & Griffin, 2006). This means that transformational leaders stimulate and transform their followers' awareness and perceptions about certain problems and offer solutions Khorshid and Pashazadeh (2014). They provide an intellectually stimulating workplace/environment, where the leaders can nurture the development of creative solutions to problems that might jeopardise the organisation's effort (Sánchez-Cardona et al., 2018). Additionally, intellectually stimulating leaders encourage followers (employees) to develop their own abilities and competences to identify, understand, and deal with future problems. Nielsen and Munir (2009, p. 315), for example suggested that 'through intellectual stimulation leaders encourage followers to make their own decisions and be creative and innovative in their work, and as such they may feel more challenged and thereby also more aroused'. As Avolio, Zhu, Koh, and Bhatia pointed out followers can creatively analyse and deal with organisational problems without direct supervision of the leader as they become innovative problem solvers themselves, which enhances improvement in their ability to rationalise issues as it relates to corporate business failures and set action plans and ability to challenge poor practices (Sánchez-Cardona et al., 2018). This means that stakeholders are prepared to deal with self-perceptions and encourage taking over more responsibility, which ultimately help them develop into leaders themselves for continued performance.

Inspirational Motivation

Motivating employees is crucial for business growth, this translates followers' (employees') skills, knowledge, and abilities into effort and performance by determining the direction and attitude of work-related behaviours and in turn leading to leading to job performance. An nspirational leader energises and empowers his/her followers by expressing a compelling vision to the organisation (Roßnagel, 2016), empowering others to contribute towards the effectiveness and success of the organization" (House et al., 1999, p. 184). This suggests that such leaders motivate employees through the instrumentality of providing solutions to employees' challenges and meaning to their work (Roßnagel, 2016). For example, such leaders can be likened to motivational speakers who build enthusiasm, team spirit, and optimism amongst team members. They engage with their followers passionately and enthusiastically about problems facing the organisation and how they can solve issues, as well as cultivate the spirit of confidence amongst team members. Therefore, inspirational motivation leaders accomplish this through one-to-one conversations and supervision, cultivating optimism teamwork, and positive outcomes. This behaviour enables employees to be more committed to their responsibilities for more productivity and working in concert with organisational goals, especially promoting corporate governance practices for a better accountable and sustained society.

Idealised Influence

This is the ability of a leader to show trust, values, and emphasise the importance of purpose, commitment, and ethical consequences of decisions whilst running daily business (Bargal, 2000). They are role models for their followers as they can be held accountable for their actions. This means that an idealised leader in an organisation is seen as role models and who can engender a sense of trust and accountability for their actions (Kotter, 1990) by articulating organisational goals and demonstrating their determination to achieve those goals (Afshari, 2020; Liden (2012). This behaviour is admired by the followers, promoting the importance of the leader–follower relationships in shaping follower commitment in different cultural contexts. For example, for business growth, it becomes crucial followers emulate emulating a leader's behaviour to developing relational commitment that enhances commitment, leading to business growth. More so, research findings of Joo et al. (2013) noted that employees/followers contribute more to positive outcomes when they perceive idealised influence of their leaders, with a sense of meaning and challenge by: sharing risks with followers, promoting the inclusive and broad-based vision and demonstrating a strong commitment to organisational goals as well as promoting and advancing accountability and transparency in the business world.

Individualised Consideration

The most significant component of transformational leadership is the leader's individualised consideration of their employees/followers. This allows the leader to become familiar with their followers' needs and circumstances, enhance communication, and improve information exchange (Rafferty & Griffin, 2006; Bass, 1985); leading to employees'/followers' success in the business. It is right for Bass to argue that individualised consideration takes two forms—negative and positive feedback, which are aimed directly at developing the employee who is expected to complete a given job role whilst learning from successes as well as mistakes (Bass, 1985). Providing individualised consideration, the leader becomes familiar with the needs of each employee/follower and mentors them (Judge & Piccolo, 2004). For instance, a leader can achieve individualised consideration through a positive influence on commitment, trust, empowerment, job satisfaction, motivation, and self-efficacy beliefs, which agree with the needs and developments of such employees (Samuel et al., 2006). This means that an employee is regarded by the leader as an individual and he/she provides developmental support that creates growth opportunities for such an individual/employee. The leader pays attention to followers' differences, discovering what motivates them, this behaviour enhances and promotes employees' skills and self-efficacy, self-fulfilment, and self-worth and, therefore, has transformational effects on business outcomes and goals.

Therefore, from the foregoing, a transformational leader recognises that staff behaviour is not just driven by standard rewards and punishment, however, they are greatly motivated by stimulating opportunities and to work together towards a common good and purpose. This means that there is less emphasis on formal authority and control, with more on collective action and commitment towards a better outcome for the people in meeting vision, values, and common purpose of the organisation. Thus, the responsibility of coalescing the ethos and translating them into actual behaviours that lead to goal accomplishment falls under the parameter of a leader who has the best interest of the stakeholders and organisation alike. It can be argued here that transformational leadership (the 4Ls) can appeal to all businesses, which is driven by change and transformation for better corporate business transparency and accountability as well as reshape corporate governance behaviour.

Prospecting for Change: Towards Transformed Corporate Behaviour and Governance in Nigeria

The recent global corporate governance scandals have redirected the thinking of regulators, policymakers, and government alike towards enacting stringent mechanisms and more robust rules to ensure good corporate governance—adequate disclosure and accountability in the corporate world (Adegbite, 2010). As outlined earlier, one of the major implications of these crises is poor leadership and inept behaviour including corruption resulting in a lack of investors' confidence. It is important for leaders (transformational) to lead the way in changing the current corporate landscape in Nigeria through their behaviour, which should inspire, motivate, and lead the way for more ethical practice as well as engender investors' confidence. Transformational leaders are responsible for a number of functions categorised as "managerial" such as setting achievable goals, communicating directives to members, overseeing execution of the organisational strategy, and assessing performance, which is linked to business growth, sustainability, and change process (Nwagbara & Ugwoji, 2015). Transformational leaders have a great influence on their followers, as they pay individual consideration to followers, thinking about various possibilities and acting self-sacrificially. These beliefs, behaviours, and values in transformational leadership can be beneficial for long-term development.

Transformational leaders instil these values and beliefs in their followers, which motivate them to learn and grow, enabling business success. This means that transformational leaders invest time and effort in developing followers because they believe that their competencies can stimulate and reshape corporate business. The followers are willing to put forth a high effort and commitment to achieve desired business goals. Research found that leadership (transformational) is central to promoting the goals/objectives of corporations (Nakpodia, 2016) as well as promoting high organisational change (Michel et al., 2011; Bcock-Roberson & Strickland, 2010; Boga & Ensari, 2009; Van Eeden et al., 2008). In prospecting for change in Nigeria's corporations, transformational leaders confront and tackle business issues including inequity transparency, accountability, corruption, and weak board of directors. To this end, this chapter argues that transformational leaders can bring about change to corporate governance through their leadership behaviour including:

- Internal motivation and self-management: Transformational leaders find expression in motivating and inspiring followers as a driving force to effectively manage an organisation to attain its common purpose. This means that a leader motivates followers by enabling them to rethink old ways of doing business, ensuring that the values of the employees align with that of the organisation, thereby promoting business success. For example, such a leader propels people to work responsibly and become accountable for their duties, thereby increasing the organisations efficiency (Van Eeden et al., 2008).
- Positive vision: Transformational leader articulates positive and inspirational vision to the employees and set a standard and explain how to reach business performance (Bass & Avolio, 2000). This enhances the vision of the business, thus increasing team-goal commitment. Transformational leaders explain to the teams' goals and performance standards to actualise change to organisations, employees, as well as foster trustworthy culture. This means that leaders constantly seek favourable means for change process in an organisation (Peng et al. (2021). They create emotional relationships with employees and inspire higher values in them. For example, such leadership conveys the importance of having a shared vision, central purpose, and direction leading to business accountability (Bass, 1999). Transformational leaders also influence board composition by

selecting directors who share their vision and values and closely with the board to ensure that the leadership style and practices align with the organisation's goals and corporate governance principles.

- Organisational consciousness: Transformational leaders are conscious and susceptible to change, and they amplify collective interest amongst employee members of an organisation to achieve collective goals (Bass & Avolio, 2000); as they seek new ways to respond to the changing business environment. This means they understand what actions to take to evoke team spirit, spur innovation, and make critical decisions that will create growth. Transformational leaders initiate/take and manage risks and take innovative approaches to growing businesses. Since their own values are aligned with the organisation, they share a joint purpose with the organisation. The leaders do not just view their position in the company as just a job, however, work towards organisational success.
- Creating a positive organisational culture: Transformational leaders create and transmit organisational innovative culture by fostering open communication, trust, and collaboration amongst employees, which in turn emphasises transparency, accountability, and ethical conduct (Bass, 1999; Bass & Avolio, 2000). This means that leaders are committed to the goals of the organisation and their positivity enables employees to seek commitment to results that promote corporate accountability and business ethical standards.

Conclusion

The importance of transformational leadership in changing the dynamics of corporate governance and behaviour cannot be overstated. The preceding argumentation lends credence to this. It creates a positive work environment, promotes ethical conduct, empowers employees, and fosters continuous learning and development. By implementing transformational leadership styles, organisations can establish effective corporate governance practices, leading to long-term success and sustainability. Both transformational leadership and corporate governance are two critical elements that shape the success and sustainability of any organisation. The chapter has explored thoroughly the significance of transformational leadership in promoting corporate governance in Nigeria and how transformational leadership can contribute to improved governance practice in Nigerian organisations.

References

Adegbite, E. (2010). *The determinants of good corporate governance: The case of Nigeria*. PhD thesis,. Case Business School, City University London.

Adegbite, E. (2012). Corporate governance regulation in Nigeria. *Corporate Governance, 12*(2), 257–276.

Adegbite, E. (2015). Good corporate governance in Nigeria: Antecedents, propositions and peculiarities. *International Business Review, 24*(2), 319–330.

Adegbite, E., Amaeshi, K., & Nakajima, C. (2013). Multiple influences on corporate governance practice in Nigeria: Agents, strategies and implications. *International Business Review, 22*(3), 524–538.

Afshari, L. (2020). Motivating toward organizational commitment: A cross-comparative perspective. *International Journal of Cross Cultural Management, 20*(2), 141–157.

Akinpelu, O. A. (2012). *Corporate governance framework in Nigeria: An international review*. iUniverse Publishers.

Asif, M., Jameel, A., Hussain, A., Hwang, J., & Sahito, N. (2019). Linking transformational leadership with nurse-assessed adverse patient outcomes and the quality of care: Assessing the role of job satisfaction and structural empowerment. *International Journal of Environmental Research and Public Health, 16*(13), 2381.

Avolio, B. J. (2007). Promoting more integrative strategies for leadership theory-building. *American Psychologist, 62*(1), 25–33.

Bargal, D. (2000). The manager as leader. In R. Patti (Ed.), *The handbook of social welfare management*. Sage.

Bass, B. M. (1999). Two decades of Research and Development in transformational leadership. *European Journal of Work and Organizational Psychology, 8*, 9–32.

Bass, B. M. (1985). Leadership: Good, better, best. *Organizational Dynamics, 13*(3), 26–40.

Bass, B. M., & Avolio, B. J. (2000). *MLQ multifactor leadership questionnaire*. Mind Garden.

Bass, B. M., & Riggio, R. E. (2006). *Transformational leadership* (2nd ed.). Psychology Press.

Bcock-Roberson, M. E., & Strickland, O. J. (2010). The relationship between charismatic leadership, work engagement, and organizational citizenship behaviour. *The Journal of Psychology, 144*(3), 313–326.

Bennis, W. (2010). Art of followership. *Leadership Excellence, 27*, 3–4.

Bennis, W. (1995). Leadership theory and administrative behavior: The problem of authority. *Administrative Science Quarterly, 4*, 259–260.

Boga, L., & Ensari, N. (2009). The role of transformational leadership and organizational change on perceived organizational success. *The Psychologist-Manager Journal, 12*(4), 235–251.

Burch, T. C., & Guarana, C. L. (2014). The comparative influences of transformational leadership and leader–member exchange on follower engagement. *Journal of Leadership Studies, 8*(3), 6–25.

Burns, J. M. (1978). *Leadership*. Harper & Row.

Cadbury, A. (2002). *Corporate governance and chairmanship: A personal view*. University Press.

Clarke, T. (2007). *International corporate governance: A comparative approach*. Routledge.

Claessens, S., & Yurtoglu, B. (2013). Corporate governance in emerging markets: A survey. *Emerging Markets Review, 15*, 1–33.

Coombes, P., & Chiu-Yin Wong, S. (2004). Why codes of governance work. *McKinsey Quarterly, 2*, 48–53.

Emeseh, E., & Songi, O. (2014). CSR, human rights abuses and sustainability report accountability. *International Journal of Law and Management, 56*(2), 136–151.

Essen, M., Engelen, P., & Carney, M. (2013). Does "Good" corporate governance help in a crisis? Corporate governance. *An International Review, 21*(3), 201–224.

Eva, N., Sendjaya, S., Prajogo, D., Cavanagh, A., & Robin, M. (2018). Creating strategic fit: Aligning servant leadership with organizational structure and strategy. *Personnel Review, 47*(1), 166–186.

Fisher, E. (2009). Motivation and leadership in social work management: A review of theories and related studies. *Administration in Social Work, 33*, 347–367.

Graen, G. B., & Uhl-Bien, M. (1995). Relationship-based approach to leadership: Development of leader–member exchange (LMX) theory of leadership over 25 years: Applying a multi-level multidomain perspective. *Leadership Quarterly, 6*, 219–247.

Hoffmeister, A., et al. (2014). The differential effects of transformational leadership facets on employee safety. *Safety Science, 62*, 69–78.

House, R., et al. (1999). Cultural influences on leadership and organizations: Project glob. In W. Mobely et al. (Eds.), *Advances in global leadership* (pp. 171–233). JAI Press.

Gao, P., & Zhang, G. (2019). Accounting manipulation, peer pressure, and internal control. *Accounting Review, 94*(1), 127–151.

Gerard, L., McMillan, J., & D'Annunzio-Green, N. (2017). Conceptualising sustainable leadership. *Industrial and Commercial Training, 49*(3), 116–126.

Gisselquist, R.M. (2012). *Good governance as a concept, and why this matters for development policy*. WIDER Working Paper No. 30. Accessed Sep 09, 2023, from http://www.wider.unu.edu/publications/working-papers/2012/en_GB/wp2012-030/

Hirschman, A. O. (1984). Exit, voice and loyalty: Responses to decline in firms, organizations, and states. *Harvard University Press, 2*, 244.

Joo, Y. J., Lim, K. Y., & Kim, J. (2013). Locus of control, self-efficacy, and task value as predictors of learning outcome in an online university context. *Computers & Education, 62*(1), 149–158.

Judge, T., & Piccolo, R. (2004). Transformational and transactional leadership: A meta-analytic test of their relative validity. *Journal of Applied Psychology, 89*(5), 755–768.

Kazemian, S., Shauri, N., Sanusi, Z., Kamaluddin, A., & Shuhidan, S. (2017). Monitoring mechanisms and financial distress of public listed companies in Malaysia. *Journal of International Studies, 10*(1), 92–109.

Khorshid, S., & Pashazadeh, A. (2014). The effect of transformational leadership on organizational learning capabilities with respect to the mediating role of organizational intelligence. *Journal of Change Management, 6*(11), 7.

Klapper, L. F., & Love, I. (2002). Corporate governance, investor protection and performance in emerging markets. *SSRN Electronic Journal, 10*, 703–728.

Kotter, J. (1990). *A force for change*. Free Press.

Kouzes, J. M., & Posner, B. Z. (2007). *The leadership challenge* (4th ed.). Jossey-Bass.

Lawler, J. (2007). Leadership in social work: A case of caveat emptor? *British Journal of Social Work, 37*(1), 123–141.

Liden, R. (2012). Leadership research in Asia: A brief assessment and suggestions for the future. *Asia Pacific Journal of Management, 29*(2), 205–212.

Mammassis, C. S., & Kostopoulos, K. C. (2019). CEO goal orientations, environmental dynamism and organizational ambidexterity: An investigation in SMEs. *European Management Journal, 37*(5), 577–588.

Michel, J. W., Lyons, B. D., & Cho, J. (2011). Is the full-range model of leadership really a full-range model of effective leader behavior? *Journal of Leadership & Organizational Studies, 18*, 1–45. Accessed Aug 09, 2023, from. https://doi.org/10.1177/1548051810377764

Nakpodia, F. (2016). *An assessment of institutional influences on corporate governance in Nigeria: A multi-stakeholder perspective*. Doctoral thesis,. Northumbria University.

Nielsen, K., & Munir, F. (2009). How do transformational leaders influence followers' affective Well-being? Exploring the mediating role of self-efficacy. *Work & Stress, 23*(4), 313–329.

Northouse, P. G. (2021). *Leadership: Theory and practice* (9th ed.). SAGE Publications.

Nwagbara, U., & Ugwoji, C. A. (2015). Corporate governance, CSR reporting and accountability: The case of Nigeria. *Economic Insights–Trends & Challenges, 67*(1), 77–84.

Nwagbara, U. (2014). *A critical discourse analysis of "managerial capture" in CSR disclosures of oil and gas companies in Nigeria*. Doctoral thesis. University of Wales.

Okike, E. N. M. (1994). Curious auditing regulations in Nigeria: A case study of cultural/political influences on auditing practices. *The International Journal of Accounting, 29*, 78–91.

Okike, E. N. M. (2007). Corporate governance in Nigeria: The status quo. *Corporate Governance, 15*(2), 173–193.

Peng, J., Li, M., Wang, Z., & Lin, Y. (2021). Transformational leadership and employees' reactions to organizational change: Evidence from a meta-analysis. *The Journal of Applied Behavioral Science, 57*(3), 369–397.

Rafferty, A. E., & Griffin, M. A. (2006). Refining individualized consideration: Distinguishing developmental leadership and supportive leadership. *Journal of Occupational and Organizational Psychology, 79*, 37–61.

Roßnagel, C. S. (2016). Leadership and motivation. In *Leadership today 2016: Practices for personal and professional performance.* Accessed July 02, 2022, from. https://doi.org/10.1007/978-3-319-31036-7_12

ROSC. (2004). *Reports on the observance of standards and codes.*

Sadeghi, A., & Pihie, Z. A. L. (2012). Transformational leadership and its predictive effects on leadership effectiveness. *International Journal of Business & Social Science, 3*, 186–197.

Samuel, N., Sleegers, P. J. C., & Denessen, E. (2006). Transformational and transactional leadership effects on teachers' job satisfaction, organizational commitment, and organizational citizenship behaviour in primary schools: The Tanzanian case, school effectiveness and school improvement. *An International Journal of Research, Policy and Practice, 17*(2), 145–177.

Sánchez-Cardona, I., Soria, M. S., & Llorens-Gumbau, S. (2018). Leadership intellectual stimulation and team learning: The mediating role of team positive affect. *Universitas Psychologica, 17*(1), 1–16.

Sergiovanni, T. J. (1984). Leadership and excellence in schooling. *Educational Leadership, 41*(5), 4–13.

Spiers, L. (2017). Corporate governance, risk and crises in small companies: Shedding light from inside the boardroom black box. *Economics and Business Review, 3*(17), 1–20.

Tafvelin, S., Hyvönen, U., & Westerberg, K. (2012). Transformational leadership in the social work context: The importance of leader continuity and co-worker support. *Journal of Social Work, 44*(4), 886–904.

Useem, M. (2001). How to groom leaders of the future. In *Financial times mastering management.* Prentice Hall.

Van Eeden, R., Cilliers, F., & Van Deventer, V. (2008). Leadership styles and associated personalities traits: Support for the conceptionalization of transactional and transformational leadership. *South Africa Journal of Psychology, 38*, 253–267.

Vanguard. *Nigeria: How country lost 75 banks to poor corporate governance*, 2007–2019. Accessed Aug 09, 2023, from https://allafrica.com/stories/200711190560.html

Wang, D., & Zhou, F. (2016). Improvement of small board listed Company's internal accounting controls. *Open Journal of Social Sciences, 4*(7), 152–158.

Wilson, I. (2006). Corporate governance in Nigeria post banking consolidation. In *ICAN@UK, newsletter of the Institute of Chartered Accountants of Nigeria (UK) District Society* (Vol. 8, pp. 1–10).

Yang, Y. U. (2015). *Earnings management and corporate governance mechanisms prior to leveraged buyouts in the UK* (Durham theses). Durham University. Available at Durham E-Theses Online http://etheses.dur.ac.uk/11512. Accessed on 01/07/2023

Yukl, G. (2006). *Leadership in organizations* (6th ed.). Pearson Education.

Chigozie A. Ugwoji, BSc, MA, MSc is a doctoral candidate at Brunel University, London, UK. She is currently a lecturer at University of Sunderland in London and is a certified social worker with many years of experience. Chigozie's areas of specialties include leadership, sustainable development, social work, management, and corporate governance.

Chapter 5
Customer Social Responsibility Reporting in Bolivia: Practices and Trends

Boris Christian Herbas-Torrico, Carlos Alejandro Arandia-Tavera, and Pamela Mirtha Zurita-Lara

Introduction

Corporate Social Responsibility (CSR) has evolved in recent years to become the focus of attention for several industries. For example, nowadays, people evaluate a firm's corporate reputation, focusing on environmental and social issues (ElAlfy et al., 2020). Therefore, visibility and communication of CSR actions and practices are substantial elements to show stakeholders and the general public about the firms' contribution to society. To this end, CSR reports were implemented, allowing for a systematic approach to managing the activities of an organization and communicating opportunities and risks in the long term (Moravcikova et al., 2015). CSR reports, unlike financial reports, allow for flexibility in the information to be shown. In other words, it is possible to present basic information, such as only mentioning the organization's activities or submitting detailed reports about environmental, economic, or social costs (Fortanier et al., 2011).

However, the emergence of CSR reports has generated mistrust about the credibility of the information presented (Michelon et al., 2015; Azapagic & Perdan, 2000). As a result, the information is evaluated according to different criteria and guidelines. For example, Michelon et al. (2015) suggest that the review should be done against sustainability standards, standardized reporting guidelines (such as the Global Reporting Initiative (GRI) indicators guide), and analyzing the management orientation given to the report. In the case of Bolivia, financial firms are required by

B. C. Herbas-Torrico (✉)
Tecnologico de Monterrey, Guadalajara, Mexico

C. A. Arandia-Tavera
Alicorp, Cochabamba, Bolivia

P. M. Zurita-Lara
YPFB Refinación S.A., Cochabamba, Bolivia

U. Nwagbara et al. (eds.), *Corporate Social Responsibility Disclosure in Developing and Emerging Economies*, CSR, Sustainability, Ethics & Governance, https://doi.org/10.1007/978-3-031-61976-2_5

Law 393 and the CSR regulation to present sustainability, and CSR reports at the end of every year (ASFI, 2014). These reports only cover financial firms, leaving other industries to do it voluntarily. The decision to only evaluate financial firms is related to the amount of money they show in their financial statements, as these reports help contrast the information assessed by a governmental office named ASFI. Consequently, this study analyzes the CSR reports presented by financial firms and CSR reports of firms from different industries and compares them with other CSR reports from international firms operating in Bolivia.

Literature Review

CSR Reporting Lawrence and Weber (2008) indicate that a firm must be responsible for all its actions and their effects on the environment and society to ensure they reduce or repair any damage that may have been caused. Gray et al. (1996) established that a corporate social report or CSR report communicates social and environmental effects to stakeholders and the general public. In other words, a CSR report can be understood as an accountability process for organizations. Reporting is a practice that allows organizations to measure, disclose, and account for their performance to internal and external authorities (GRI, 2006). Therefore, these reports include key performance indicators for environmental, economic, and social performance (Bonsón & Bednárová, 2015). Due to stakeholders' pressure to know the environmental and social governance indicators, firms have improved transparency when reporting their performance using internationally accepted standards, such as the Global Reporting Initiative standard (Arvidsson, 2010). Moreover, other reasons why firms do CSR reports are: (a) to show their responsibility to all the stakeholders (Deegan & Samkin, 2006); (b) to answer all the expectations and contribute to the well-being of society (Morsing & Schultz, 2006); (c) to protect corporate identity and reputation in front of their stakeholders (Reynolds & Yuthas, 2008); and (d) to assure long-term profitability by reducing asymmetries in information and improving decision making (Merkl-Davies & Brennan, 2007; Du et al., 2010).

GRI Standards The Global Reporting Initiative (GRI) is an independent organization that focuses on helping organizations take action on their activities by communicating standards in numerous dimensions (GRI, 2006). These standards are designed for organizations to report on multiple business aspects, such as waste management or health and safety in the workplace. These standards are grouped into different dimensions to show broader environmental, economic, or social themes. According to Li and Toppinen (2011), grouped GRI standards allow a global view of a firm's effects. The groups are as follows:

1. *Economic responsibility*. This group has the following factors: (a) economic performance, (b) market presence, and (c) indirect economic impacts.

2. *Labor practices responsibility*. This group covers (a) employment, (b) labor relations, (c) health and safety in the workplace, (d) education and training, and (e) equal opportunities and diversity.
3. *Social responsibility*. This group includes (a) community, (b) corruption, (c) public policies, (d) anti-competitive behavior, and (e) compliance.
4. *Environmental responsibility*. This group takes into account: (a) materials, (b) energy, (c) water, (d) biodiversity, (e) waste, effluents, and emissions, (f) products and services, (g) compliance, (h) transport, and (i) overall.
5. *Human rights responsibility*. This group has the following factors: (a) investment and procurement, (b) non-discrimination, (c) freedom of association, (d) child labor, (e) forced labor, (f) security practices, and (h) indigenous rights.
6. *Product responsibility*. This group considers (a) consumer safety, (b) product labeling, (c) marketing communications, (d) consumer privacy, and (e) compliance.

Bolivian CSR Reporting Law In Bolivia, compliance with CSR is regulated by many laws. However, in this study, we focus mainly on two regulations. One is the Law 393 of financial services, which aims to control the activities of financial intermediation and the provision of financial services, as well as the organization and operation of economic entities (ASFI, 2013). In particular, this law requires financial institutions to submit financial, social activities, and compliance reports to understand how they move their resources and for what. This law also reviews whether financial firms offer equal opportunities and transparent information for their clients and employees.

On the other hand, we also focus on the Management of Corporate Social Responsibility regulation developed by a governmental institution named ASFI (2014). This regulation requires financial institutions to report on accountability, transparency, ethical behavior, stakeholder engagement, compliance with laws and regulations, and respect for human rights in their CSR reports. In summary, this norm gives firms the steps to prepare their reports and highlight the use of the GRI standard. Consequently, this regulation makes financial firms focus on the standards related to the economy, environment, human rights, community, and social activities to show their stakeholders an overview of their global activities and their contribution to society.

Besides both regulations, there are other public documents with suggestions to apply or implement CSR activities voluntarily. The most common is the Corporate Social Responsibility Guide elaborated by the Ministry of Productive Development and Plural Economy in 2018. This guide is adopted voluntarily for its use or implementation and only contains suggestions for developing a social responsibility system, not detailed information to be presented to stakeholders. In summary, it tells firms how to establish their CSR system and does not give them the insights to share this information with their stakeholders. Therefore, we did not consider it relevant to our study.

Methodology

To conduct our study about CSR reporting in Bolivia, we downloaded CSR reports from ten financial institutions, ten manufacturing and service firms, and ten foreign firms with operations in the country. Once the reports were obtained, we proceeded to analyze their structure. This analysis aimed to identify whether they follow GRI standards and quantify the number of reported dimensions. Moreover, we also grouped them based on Li and Toppinen (2011). In particular, this was done to determine which dimensions firms are more inclined to fulfill, considering whether they are under mandatory CSR laws or other regulations. In Table 5.1, we can see that the environmental responsibility standards (90.32%), societal responsibility (87.10%), and human rights responsibility (83.87%) have the highest compliance, whereas product responsibility (19.35%), economic responsibility (70.97%), and labor practices responsibility are the least complied (77.42%).

Results

Financial firms The financial industry firms' results indicate that environmental responsibility standards have the highest level of compliance, appearing in 100% of CSR reports. This increased compliance may be due to their relationship with the legal requirements outlined in the ASFI CSR regulation (2013). Economic and human rights responsibility standards also appear in high numbers, with 90.91% of reports showing compliance, which is also related to the ASFI's CSR regulation. This regulation requires financial firms to demonstrate their fulfillment of environmental and human rights requirements for themselves and their clients. The societal responsibility and labor practices responsibility standards are in third and fourth place, with 81.82% and 63.64% compliance, respectively. These standards are impacted by some financial firms' partial reporting of activities related to them. Finally, the product responsibility standard has the lowest level of compliance, appearing in only 9.09% of reports. This low level of compliance is because the products provided by financial firms, such as electronic transactions and account openings, are not 100% physical. Hence, the focus of their CSR efforts is on economically oriented service standards and not on physically palpable product

Table 5.1 General compliance with GRI standards

GRI dimension	Number of firms	Percentage
Economic responsibility	22	70,97%
Environmental responsibility	28	90,32%
Labor practices responsibility	24	77,42%
Human rights responsibility	26	83,87%
Societal responsibility	27	87,10%
Product responsibility	6	19,35%

standards (Li & Toppinen, 2011). Moreover, we found that these reports follow the GRI standard structure. A summary of the collected information is presented in Table 5.2.

Manufacturing and service firms The results obtained for Bolivian manufacturing and service firms show that societal responsibility standards are the most prevalent (100%) in CSR reports. This result occurs because firms aim to contribute to their community by repairing public spaces or lighting (Davis & Franks, 2011). The labor practices responsibility and human rights responsibility standards are shown in 88.89% of the CSR reports. This result suggests that Bolivian firms follow the Bolivian Labor Law and the CSR guide of the Ministry of Productive Development and Plural Economy (2018) when developing their policies regarding human rights and working practices. For instance, they comply with requirements such as social security and holiday pay benefits from the labor law and implement safety measures and training sessions for their employees, as in the CSR guide. Turoń (2016) suggests that good working practices boost staff motivation and improve firm results. The environmental and economic responsibility standards appear in 77.78% of the CSR reports due to the difficulty in being transparent about environmental and financial information (Michelon et al., 2015) in Bolivia. In particular, Bolivian firms avoid publishing such information due to the lack of support to justify or clarify any potential non-compliance with regulations or expectations. Finally, the standard of product responsibility is the lowest at 11.11%. This result occurs because, according to GRI (2006), product responsibility is linked to information and business activities aimed at taking care of consumers. However, firms cannot ensure compliance with all regulations, such as labeling legislation, due to the prolonged duration of administrative procedures to modify labels in Bolivia. This issue is also evident in evaluating suppliers, which is complex due to the high number of informal suppliers in Bolivian supply chains (Morales, 2020). Table 5.3 provides a summary of these results. Moreover, these CSR reports do not fully follow the GRI standards guide, as the sequence and identification of sectors to focus on are based on the firm's situation.

International manufacturing and service firms In the case of global manufacturing and service firms operating in Bolivia, our results indicate that environmental responsibility standards (76.92%) are the most prevalent in CSR reports (see

Table 5.2 GRI dimensions compliance for financial firms

GRI dimension	Number of firms	Percentage
Economic responsibility	10	90,91%
Environmental responsibility	11	100,00%
Labor practices responsibility	7	63,64%
Human rights responsibility	10	90,91%
Societal responsibility	9	81,82%
Product responsibility	1	9,09%

Table 5.4) because firms often prioritize environmental concerns, as they are scrutinized by customers (Puppim de Oliveira & Jabbour, 2017). The second most common standards are labor practices responsibility and societal responsibility (69.23%). These standards receive this classification because they relate to two aspects: (i) showing good labor practices to retain or attract new talent and improve the firm's image (Turoń, 2016) and (ii) demonstrating to society at large the contributions made by the firm, such as donations or awareness campaigns, to show their collaborative role in society (Chae et al., 2014). Human rights responsibility standards (61.54%) are in third place, focusing mainly on compliance with human rights policies in core business operations and support for projects promoting good human rights practices, such as supporting UNICEF campaigns (Jadhav & Aswale, 2012). In fourth place are economic responsibility standards (38.46%), and in last place are product responsibility standards (30.77%), as shown in Table 5.4. These two standards have low scores as they can be prone to confusion. Davis and Franks (2011) note that the goal is to connect with the community through transparent and reliable information, so if there is no accurate information to justify or explain a situation, it is best to avoid publishing info that may cause confusion or distrust. Finally, not all analyzed reports follow the format suggested by GRI standards, leading to greater dispersal in the standards used.

GRI standards per firm type Regarding the firm type, Fig. 5.1 shows that economic responsibility standards are widely used by the financial sector, with a usage rate of 90.91%. On the other hand, international firms give a lower usage rate of these standards at 38.46%. This difference can be explained by the findings of Karaman et al. (2021), who suggest that economic measures help secure opportunities for development or financing, making them a more common practice in emerging economies. In other words, the need for economic growth in these countries may drive firms to present GRI standards of this type to attract international funding.

Regarding environmental responsibility standards, financial firms report 100% on this dimension, while firms (national and international) only report 77%. Usually, the environmental responsibility dimension is the most worked on in CSR in Latin America (Halkos & Nomikos, 2021). However, this report has begun to generate distrust as it hides or camouflages possible violations (Ordóñez-Castaño et al., 2021; Moneva et al., 2006). As a result, Ordóñez-Castaño et al. (2021) suggest that these standards should not only present performance in CSR but also actions to

Table 5.3 GRI dimensions compliance for Bolivian manufacturing and service firms

GRI dimensions	Number of firms	Percentage
Economic responsibility	7	77,78%
Environmental responsibility	7	77,78%
Labor practices responsibility	8	88,89%
Human rights responsibility	8	88,89%
Societal responsibility	9	100,00%
Product responsibility	1	11,11%

Table 5.4 GRI dimensions compliance for international manufacturing and service firms

GRI dimension	Number of firms	Percentage
Economic responsibility	5	38,46%
Environmental responsibility	10	76,92%
Labor practices responsibility	9	69,23%
Human rights responsibility	8	61,54%
Societal responsibility	9	69,23%
Product responsibility	4	30,77%

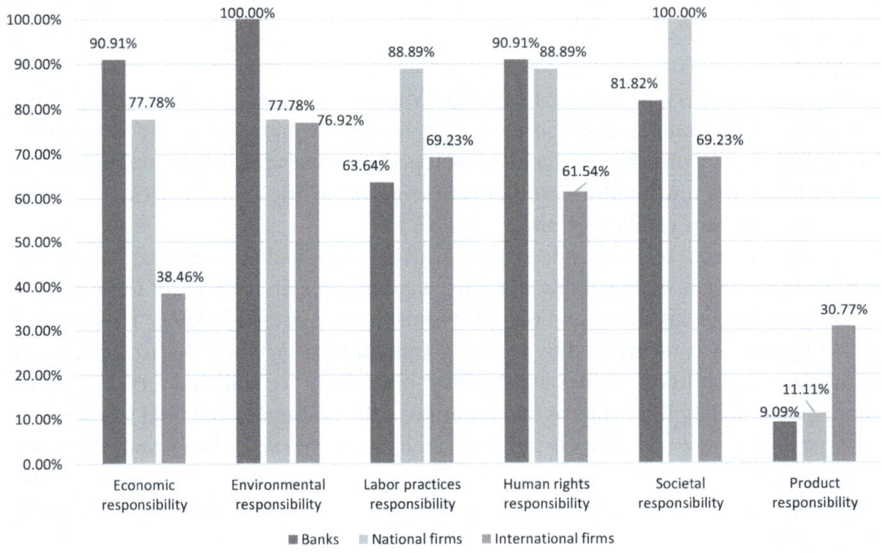

Fig. 5.1 GRI Standards per firm type

correct observations and show the progress made to improve weak points to be transparent with the information published. In this way, compliance with environmental standards will not seek the highest compliance but will seek to show the best opportunities for transparency in communication (Moneva et al., 2006).

Regarding labor practices responsibility standards, national firms lead in the appearance of this standard with 88%, while other sectors are around 68%. This result is explained by Turoń (2016), who suggests that while developed countries have difficulties improving their performance, developing countries can obtain better results. Moreover, developing countries often do not regularly comply with correct labor practices due to the high number of informal jobs (Bacchetta & Bustamante, 2009).

For human rights responsibility standards, we can see that the financial sector and national firms have the highest compliance, with approximately 89%. At the same time, international firms show this standard in 61% of their reports. At this point, something similar happens, as mentioned in labor practices responsibility standards. Developing countries tend to have a low implementation of these standards, reflected in a lower appearance of these standards in reports because, compared to developed countries, developing countries tend to have weak institutions (Wettstein, 2012). However, as Perello-Marin et al. (2022) noted, the approach to follow is to make human rights a part of a firm's KPIs in Bolivian firms.

According to the data on the standard of societal responsibility, we can see that again, the financial and national firms are the ones that report the highest appearance with an average of 90%, while international firms only show 69%. According to Halkos and Nomikos (2021), Latin America and the Caribbean had the most take-off in using these standards. This includes human rights and the relationship of this with inequalities. Thus, it is more common for human rights reporting standards to be higher in Latin American countries because there is a more significant social gap to improve. Whereas, according to Perello-Marin et al. (2022), adopting the report of human rights standards in Europe tends to require more analysis, where only pending gaps are considered. Hence, developed countries present compliance standards related to a few pending problems to work on, while Latin America is reporting significant gaps (Halkos & Nomikos, 2021).

Finally, we can see that international firms stand out in the case of product responsibility because they show that this standard appears in 30% of their reports. While in the financial sector and national firms, it only occurs in 10%. This result is related to what was mentioned by Li and Toppinen (2011), who refer to the fact that product responsibility variables are the most interesting for stakeholders. For example, responsible labeling. Halkos and Nomikos (2021) suggest that GRI standards of this type are the most complex to achieve since organizations are constantly exposed to errors or unexpected events that affect their credibility. For example, the Volkswagen scandal when the modification of emissions test results was admitted.

Conclusion and Discussion

Through this study, we could identify that GRI standards help firms report on their environmental, social, and economic performance. In the financial sector, the study found that structured and guided by GRI standards, CSR reports are presented the most. This way of working allows financial firms to comply with legal requirements such as CSR regulations and the Law 393 of the ASFI. This has allowed financial firms to present many GRI standards in their reports, especially in the environmental, human rights, and economic responsibility dimensions. On the other hand, Bolivian services and manufacturing firms, do not follow a complete report of the GRI standards but still maintain a high reporting of standards, especially in society, human rights, and labor practices responsibility dimensions. It is important to note that the presentation of these reports for these sectors is entirely voluntary. Finally, in reports from international manufacturing and services firms, we can see that they focus mainly on the environment, labor practices, and societal responsibility dimensions. It is noteworthy that in Bolivia, a common point of interest is presenting standards on human rights. Also, CSR reports report more standards in Bolivia's manufacturing and services sector than in the financial industry, even though the latter is regulated.

Additionally, the banking sector agrees with international firms in the high presentation of standards focused on the environment. Furthermore, the manufacturing and services sectors, both in Bolivia and internationally, concur in an increased reporting of standards on labor practices and society. Also, none of the studied sectors present a high degree of presentation of product responsibility standards.

In general, using GRI standards in different sectors is essential for firms and stakeholders, but it can vary among industries. This study found that financial firms present CSR reports in a structured and guided manner based on the GRI standards. On the other hand, Bolivia's manufacturing and services sector has high reporting of standards, especially in societal, human rights, and labor practices responsibility dimensions. Moreover, international firms focus mainly on environmental, labor practices, and societal responsibility dimensions.

The common point in Bolivian firms is that they show many gaps in different dimensions but compliance with human rights that reflect equal job opportunities (Turoń, 2016). On the other hand, the manufacturing and service sectors have many standards implemented despite not being regulated. Also, we found that, in general, the three most widely used standards are: (a) environmental responsibility, (b) human rights responsibility, and (c) labor practices responsibility. This result shows that despite firms working in different regions and industries, the focus of the reports is similar. Perello-Marin et al. (2022) mention that the goal of CSR reports is to present the firm's actions to contribute to stakeholders' interests, regardless of the region. Similarly, we can infer that the gaps between developed and developing countries have common points. Finally, we found that none of the industries have been able to raise the presentation of standards on product responsibility due to governmental regulations and bureaucracy.

In summary, we found that the structure of reports and the legal regulations do not incentivize CSR reporting. Therefore, the following recommendations can be made for Bolivian firms: (a) adopt a format or establish one of their own for the presentation of CSR reports, (b) divide the standards to be presented between those that are regulated and those that are voluntary, to show that the business initiative goes beyond the legal regulation of Bolivia, and (c) present reports with a past-present-future vision so that the reader can know the changes made over time and where the CSR initiatives are aimed.

No research is perfect, and ours is far from it. Future research can account for our limitations by adding more firms and variables to the analysis. For example, future research can consider implementing reports from public and informal firms. Moreover, other studies can analyze the influence of CSR reports on stakeholders and their relationship with purchasing intentions.

References

Arvidsson, S. (2010). Communication of corporate social responsibility: A study of the views of management teams in large firms. *Journal of Business Ethics, 96*, 339–354.

Azapagic, A., & Perdan, S. (2000). Indicators of sustainable development for industry: A general framework. *Process Safety and Environmental Protection, 78*(4), 243–261.

ASFI. (2013). *Financial services laws*. https://servdmzw.asfi.gob.bo/circular/leyes/Ley393ServiciosFinancieros.pdf

ASFI. (2014. *Corporate social responsibility regulation*. https://servdmzw.asfi.gob.bo/circular/textos/L03T11.pdf

Bacchetta, M., & Bustamante, J. P. (2009). *Globalization and informal jobs in developing countries: A joint study of the International Labour Office and the Secretariat of the World Trade Organization*.

Bonsón, E., & Bednárová, M. (2015). CSR reporting practices of Eurozone firms. *Revista de Contabilidad, 18*(2), 182–193.

Chae, B. K., Olson, D., & Sheu, C. (2014). The impact of supply chain analytics on operational performance: A resource-based view. *International Journal of Production Research, 52*(16), 4695–4710. https://doi.org/10.1080/00207543.2013.861616

Davis, R., & Franks, D. M. (2011, October). The costs of conflict with local communities in the extractive industry. In *Proceedings of the first international seminar on social responsibility in mining* (Vol. 30, pp. 7576–7581).

Deegan, C., & Samkin, G. (2006). *New Zealand financial accounting*. McGraw-Hill.

Du, S., Bhattacharya, C., & Sen, S. (2010). Maximizing business returns to corporate social responsibilities: The role of CSR communication. *International Journal of Management Reviews, 12*(1), 8–19.

ElAlfy, A., Palaschuk, N., El-Bassiouny, D., Wilson, J., & Weber, O. (2020). Scoping the evolution of corporate social responsibility (CSR) research in the sustainable development goals (SDGs) era. *Sustainability., 12*(14), 5544. https://doi.org/10.3390/su12145544

Fortanier, F., Kolk, A., & Pinkse, J. (2011). Harmonization in CSR reporting. *Management International Review, 51*(5), 665–696.

Gray, R., Owen, D., & Adams, C. (1996). *Accounting & accountability: Changes and challenges in corporate social and environmental reporting*. Prentice-Hall.

GRI. (2006). *RG: Sustainability reporting guidelines, global reporting initiative*. Accessed Nov 15, 2008, from www.globalreporting.org/NR/rdonlyres/ED9E9B36-AB54-4DE1-BFF25F735235CA44/0/G3_GuidelinesENU.pdf

Halkos, G., & Nomikos, S. (2021). Corporate social responsibility: Trends in global reporting initiative standards. *Economic Analysis and Policy, 69*, 106–117.

Jadhav, A. B., & Aswale, S. (2012). CSR: Gizmo pro human rights (a case of Indian multinationals practices). *Asia Pacific Journal of Management & Entrepreneurship Research, 1*(1), 100.

Karaman, A. S., Orazalin, N., Uyar, A., & Shahbaz, M. (2021). CSR achievement, reporting, and assurance in the energy sector: Does economic development matter? *Energy Policy, 149*, 112007.

Lawrence, A. T., & Weber, J. (2008). *Business and society: Stakeholders, ethics, public policy* (12th ed.). McGraw-Hill.

Li, N., & Toppinen, A. (2011). Corporate responsibility and sustainable competitive advantage in forest-based industry: Complementary or conflicting goals? *Forest Policy and Economics, 13*(2), 113–123.

Merkl-Davies, D., & Brennan, N. (2007). Discretionary disclosure strategies in corporate narratives: Incremental information or impression management? *Journalof Accounting Literature, 26*, 116–196.

Michelon, G., Pilonato, S., & Ricceri, F. (2015). CSR reporting practices and the quality of disclosure: An empirical analysis. *Critical Perspectives on Accounting, 33*, 59–78.

Ministry of Productive Development and Plural Economy (2018). *Corporate social responsibility (CSR)*. Corporate social responsibility guide. https://produccion.gob.bo/wp-content/uploads/2021/04/Guia-RSE-2019.pdf

Moneva, J. M., Archel, P., & Correa, C. (2006). GRI and the camouflaging of corporate unsustainability. In *Accounting forum* (Vol. 30., No. 2, pp. 121–137). No longer published by Elsevier.

Morales, R. (2020). The resilience of informal workers to COVID19 and to the difficulties of trade. *Sociology International Journal, 4*(4), 92–95. Retrieved from https://www.medcrave.org/index.php/SIJ/article/view/18878

Moravcikova, K., Stefanikova, Ľ., & Rypakova, M. (2015). CSR reporting as an important tool of CSR communication. *Procedia Economics and finance, 26*, 332–338.

Morsing, M., & Schultz, M. (2006). Corporate social responsibility communication: Stakeholder information, response and involvement strategies. *Business Ethics: A European Review, 15*(4), 323–338.

Ordóñez-Castaño, I. A., Herrera-Rodríguez, E. E., Franco Ricaurte, A. M., & Perdomo Mejía, L. E. (2021). Voluntary disclosure of GRI and CSR environmental criteria in Colombian firms. *Sustainability, 13*(10), 5405.

Perello-Marin, M. R., Rodríguez-Rodríguez, R., & Alfaro-Saiz, J. J. (2022). Analysing GRI reports for the disclosure of SDG contribution in European car manufacturers. *Technological Forecasting and Social Change, 181*, 121744.

Puppim de Oliveira, J. A., & Jabbour, C. J. C. (2017). Environmental management, climate change, CSR, and governance in clusters of small firms in developing countries: Toward an integrated analytical framework. *Business & Society, 56*(1), 130–151.

Reynolds, M., & Yuthas, K. (2008). Moral discourse and corporate social responsibility reporting. *Journal of Business Ethics, 78*(1–2), 47–64.

Turoń, K. (2016). Corporate social responsibility to employees: The best labour practices in transport and logistics firms. *Journal of Corporate Responsibility and Leadership, 3*(1), 37–47.

Wettstein, F. (2012). CSR and the debate on business and human rights: Bridging the great divide. *Business Ethics Quarterly, 22*(4), 739–770.

Boris Christian Herbas-Torrico is a Professor at the Tecnologico de Monterrey in Mexico. He holds a Ph.D. (D.Eng.) and a Master's degree (M.Eng.) in Industrial Engineering and Management from the distinguished Tokyo Institute of Technology in Japan. With a passion for academia and research, his expertise spans various fields, including marketing, production, applied statistics, and corporate social responsibility

Carlos Alejandro Arandia-Tavera is a production coordinator in Alicorp Bolivia. He holds a bachelor's degree in industrial engineering from the Bolivian Catholic University. He had different positions in manufacturing firms in Bolivia. He is currently pursuing a master's degree in data analytics at INCAE Business School. He had different positions in manufacturing firms in Bolivia.

Pamela Mirtha Zurita-Lara holds a bachelor's degree in industrial engineering from the Bolivian Catholic University. Her work experience is in supply chains and human resources. She currently serves as a Human talent development analyst at the YPFB Refinacion S.A. in Cochabamba, Bolivia.

Part II
Corporate Social Responsibility Governance and Sustainable Development

CSR as a Solution for Tackling Socio-economic Challenges in Romania
Silvia PUIU

Unpacking Corporate Social Responsibility Activities in Organisations by Considering Social, Economic, and Environmental Challenges: A Conceptual Approach
Afzal Munna, Md Sadeque Imam Shaikh, Uzoechi Nwagbara

Advancing Governance Towards Sustainable Development Based on Multilateralism and Democracy: The Example of Chile
Ursula A. Vavrik

Contribution of Corporate Social Responsibility (CSR) and Sustainability: Practices of Small and Medium Sized Enterprises (SME's) to Social Development in Nigeria: A Critical Assessment of Practices, Barriers and Strategies
Adebimpe Adesua and Brendhain Diamond

Corporate Social Responsibility (CSR) and Sustainable Development: Exploring the Nexus in the Sierra Leonean Context
Henry Yamba Kamara, Uzoechi Nwagbara and Ngozi Ibeawuchi

An Analysis of Corporate Social Responsibility (CSR) Related Objectives Enshrined in the 'National Oil and Gas Policy for Uganda, 2008:' A Policy Perspective Paper
David Katamba, Muhammed Ngoma, Bosco Amerit, Maureen Basuuta and Swithern, B. Tumwine

Chapter 6
CSR as a Solution for Tackling Socio-Economic Challenges in Romania

Silvia Puiu

Introduction

Corporate Social Responsibility (CSR) is extensively debated in the professional literature, being an important topic of discussion for both the public and the private sectors in an economy. During the last few years, the entire world faced numerous challenges, such as the COVID-19 pandemic, a war, an energy crisis, a rise in inflation, the interest and unemployment rates. Taking into account all these major problems combined, the role of CSR became more crucial than before.

Each crisis could be seen as a challenge for all actors involved (public authorities, private companies, and citizens) to take action and implement the appropriate solutions for solving the problems in the community. Zhou and Ki (2018) highlight the "destructive power of crises" which might affect negatively the image of a company, especially if and when CSR is not managed properly. Ursic and Cestar (2022) analyzed the difference between companies in Slovenia during the COVID-19 pandemic and noticed that those with international activity were more proactive than domestic businesses. The explanation consists of the dimension and the financial power of companies operating beyond the national borders but also in their wider perspective on things and a deeper understanding of the role played by CSR in their future.

Cho et al. (2019) found a positive correlation between the CSR involvement of Korean companies and their financial results. This finding is a great motivational factor for companies to invest in CSR programs in order to ensure their survival in a competing world. Other authors (Rhou et al., 2016) emphasize the positive relationship between CSR and profits but also highlight the role played by the awareness of the shareholders and managers of the company toward these initiatives. If

S. Puiu (✉)
University of Craiova, Craiova, Romania

U. Nwagbara et al. (eds.), *Corporate Social Responsibility Disclosure in Developing and Emerging Economies*, CSR, Sustainability, Ethics & Governance, https://doi.org/10.1007/978-3-031-61976-2_6

they do not see or understand the importance of CSR for their business, they will not reap the maximum benefits.

As Waworuntu et al. (2014) show in their study on ASEAN countries, there is a positive correlation between CSR and financial performance but this is not identical for all countries. They noticed that the correlation was weaker for countries like Malaysia and Thailand. This finding shows that it is important to understand that there are other factors affecting this relation, such as the awareness toward CSR in a culture and economy but also the level of economic development of the country and of each company that might have or not the financial resources to invest in CSR projects. Puiu and Wiśniewski (2020) explain the difference between Small and Medium Enterprises (SMEs) and multinational companies in terms of their budgets and their power to engage in big CSR programs. However, as the authors state, the benefits of CSR are important also for smaller companies.

The impact of CSR on a business and especially in challenging times like a crisis could be understood from the perspective of the consumers. There are several trends in consumer behavior that should motivate businesses to be more proactive and be more present in the community and initiate CSR projects. There are many authors (Lee & Shin, 2010; Ali & Sohail, 2018; Sharma et al., 2018; Gupta et al., 2021) who studied the relationship between CSR and the intention of consumers to buy the products and services they need from companies more oriented to being more socially responsible.

Lee and Shin (2010) explain that this relation is mediated by the consumers' awareness regarding the importance of CSR for a community. Sharma et al. (2018) appreciate the importance of morality as a motivating factor for CSR activities in the eyes of consumers. Thus, if consumers are convinced by the company that its CSR projects are motivated by morality and the desire to do good, consumers are more inclined to purchase from that company.

Another trend noticed during the COVID-19 pandemic was related to the tendency of consumers to move toward local small businesses in their neighborhood (Li et al., 2020; Danciu, 2021; Cambefort, 2020; Mehta et al., 2020) at least in 2020 and 2021. Besides the focus on local companies, Danciu (2021) also mentions the importance given by Romanian consumers to products that are more ecofriendly and this is related to the CSR initiatives of a company focused on the well-being of a community and on reducing the carbon footprint on the environment.

As well as Danciu (2021), Cambefort (2020) also noticed that consumers became more aware of the fact that consumerism affects the environment and thus, people reduced their unnecessary expenses. Mehta et al. (2020, p. 295) state that people chose more local businesses for doing their shopping, considering this a "more sustainable option." These changes in consumer behavior are important for companies because they can change the way they do business, the way they promote themselves, and the way they communicate about their CSR projects.

The health crisis generated by a global pandemic came with many other challenges: people lost their jobs and were not able to recover financially, prices increased, and the cost of living became much higher also because of the energy crisis.

The goal of the present chapter is to change the perspective from crisis to challenge to opportunity. In times of important challenges and crises on multiple levels for everyone, companies have the opportunity to be more present in the community, to help, to make a significant impact, and listen to what their consumers need most.

Thus, the structure of the present chapter includes a presentation of the methodology used, mainly the PRIMA method, followed by a presentation of results and a discussion of them, and in the end, we presented the conclusions and recommendations to raise the number of CSR initiatives that can help in solving the socio-economic problems the community faces, especially during crises.

Research Methodology

The main aim of the present paper is to identify all relevant research regarding the CSR initiatives of both the public and private sectors in Romania. Thus, the main objectives of the research are expressed in the following research questions (RQ1–6):

RQ1: What is the status quo of CSR in Romania?

The answer to this question will help with reaching the objective of identifying a pattern that might be relevant to the problems that are most stringent in this developing country.

RQ2: What are the most important problems faced by the companies developing CSR projects in Romania?

This question will help with reaching the objective of identifying the barriers that might prevent companies from being more socially responsible.

RQ3: What is the level of CSR in the Romanian public sector?

This question will help with reaching the objective of creating a more accurate image of CSR in the public sector which is not so common, neither in practice nor in the literature.

RQ4: What is the level of awareness regarding CSR and the socio-economic challenges that CSR might help within Romania?

This question will help with reaching the objective of identifying the problems related to the awareness in the community which might not encourage the companies to invest in more relevant CSR projects.

RQ5: What is the relation between CSR strategies and the financial performance of a company?

This question will help with reaching the objective of understanding the correlation between CSR and the financial results of a company due to its impact on consumers and other stakeholders.

RQ6: What happened during the economic crises with CSR initiatives in Romania and the budgets allocated for these initiatives by companies?

Answering this question will help with reaching the objective of understanding the response of companies to crises in Romania. The answer might also depend on the company size, which will be researched in the literature review.

The present study is useful for creating a proper legislative background that will stimulate more private initiatives related to CSR, especially those focused on solving some of the socio-economic challenges which the community faces in Romania. Also, it might help companies, especially small and medium enterprises to understand the benefits of their CSR projects for both their communities and their financial performance.

The method used for finding the answers to these questions and thus reaching the research objectives is the Preferred Reporting Items for Systematic Reviews and Meta-Analyses (PRISMA) (Page et al., 2021). For conducting scientific qualitative research, we used the Scopus database during February and March 2023 using specific terms in the search bar, focusing on research published between 2008 and 2022. All keywords used for searching the database are presented in Table 6.1.

The three stages of the PRISMA method (Page et al., 2021) include the identification of the papers in the literature review; the screening where the researchers can choose some criteria for including or excluding papers from the research; and the final analysis where the researchers make the analysis of the papers remained after the screening.

Table 6.2 summarizes the results for each stage of the PRISMA method. Thus, in the first stage, we searched the Scopus database using the combination of keywords in Table 6.1 and got a total of 161 results. In the second stage of screening, we used as exclusion criteria the language (we aimed only for papers in English), the status of open access, the type of document (articles), and the publication date (2008–2022).

Results and Discussions

For answering RQ1, we searched the Scopus database and got 117 results after searching the keyword *CSR in Romania*. We applied the exclusion criteria of time, language, and type, and remained with 28 papers, thus excluding 89 documents. Seventy-five percent of the articles in our analysis were published in the last 5 years (2018–2022). Most papers (12 of the 28) were published in the journal Sustainability. The first three more cited papers were also published in Sustainability:

Table 6.1 Keywords used to search the Scopus database for CSR in Romania

Research questions (RQ)	*Topic*	*Keywords used for search*
RQ1	Status quo of CSR in Romania	CSR in Romania
RQ2	Problems/barriers faced by companies that initiate CSR projects in Romania	CSR obstacles Romania / CSR barriers Romania
RQ3	CSR in the Romanian public sector	CSR public sector Romania
RQ4	Awareness regarding CSR in Romania	CSR awareness Romania
RQ5	The relation between CSR and financial performance	CSR financial performance Romania
RQ6	CSR budgets in times of crises in Romania	CSR crisis Romania

Table 6.2 The three stages of the PRISMA method

Identification	Number of papers on the Scopus database using the search criteria from Table 6.1 ($n = 161$)	Papers excluded
Screening	Number of articles in English, open access, published between 2008 and 2022 ($n = 28$). Duplicates and results not meeting the criteria were excluded.	133
Final analysis	Number of papers included in the final analysis ($n = 28$)	–

Burlea-Schiopoiu and Mihai (2019) with 38 citations, Hategan and Curea-Pitorac (2017) with 37 citations and Badulescu et al. (2018) with 33 citations. A summary of the articles included in our analysis is presented in Table 6.3.

We notice a few themes that were approached by the authors in relation to CSR in Romania: the positive correlation between CSR engagement and the companies' financial performance (Burlea-Schiopoiu & Mihai, 2019; Hategan & Curea-Pitorac, 2017; Simionescu & Dumitrescu, 2018; Matei et al., 2021; Man et al., 2021); the role of the manager in the CSR engagement and reporting (Socoliuc et al., 2018; Saveanu et al., 2021; Păun et al., 2020); different perspectives on CSR depending on the size and age of the company (Badulescu et al., 2018; Man et al., 2021; Dura & Driga, 2017), the consumers perspective on CSR and the impact on their loyalty toward brands and companies (Moisescu, 2015; Moisescu & Gică, 2017; Moisescu et al., 2019; Dabija et al., 2022), the endogenous dimension of CSR and especially the focus on companies' employees (Obrad & Gherheş, 2018; Stoian & Zaharia, 2012; Păun et al., 2020).

For answering RQ2, we searched the Scopus database and got six results after searching the keywords *CSR obstacles Romania* (four results) and *CSR barriers Romania* (two results). After eliminating a duplicate paper that had been found in both searches, we did the screening for the criteria of language, time, and the open-access status of the papers, we remained with only two papers: Facilitators of and barriers to sustainable development in small and medium-sized enterprises: *Facilitators of and barriers to sustainable development in small and medium-sized enterprises: A descriptive exploratory study in Romania* (Costache et al., 2021) and *Analysis of social responsibility and reporting methods of Romanian companies in the countries of the European Union* (Socoliuc et al., 2018). Both were also identified in the analysis for answering RQ1 so we will consider them for RQ2 but we will keep them once in the calculus in Table 6.2. The two articles were published in the journal Sustainability (indexed in the Web of Science database) and are rather recent (2018 and 2021).

Costache et al. (2021) conducted qualitative research on Romanian SMEs to identify the main barriers for them in initiating CSR programs. The authors mention "the legal and institutional environment, …unpredictable and unstimulating regulations, bureaucracy, and corruption"; not finding suppliers and customers that appreciate CSR and more sustainable business practices which impacts their decisions regarding CSR expenses; limited financial resources; not fully understanding the role the managers of the SMEs play in the CSR in Romania (Costache et al., 2021).

Table 6.3 Summary of the articles regarding RQ1

Article	*Main ideas*
An integrated framework on the sustainability of SMEs (Burlea-Schiopoiu & Mihai, 2019)	CSR developed by SMEs, positive correlation between CSR expenses and profitability, focus on construction, wholesale and retail sectors in Oltenia region
Testing the correlations between corporate giving, performance and company value (Hategan & Curea-Pitorac, 2017)	The authors conclude that there is a strong relationship between the decision to initiate CSR programs and the performance of a company. Also, companies that have more financial resources and a higher market value are more prone to allocate resources to CSR
The relationship between firm size and age, and its social responsibility actions Focus in a developing country (Romania) (Badulescu et al., 2018)	The authors make a distinction between newer and older companies in the way they develop their CSR strategies. Thus, 69.23% of newer companies are focused on donations and sponsorships, whereas older ones use the CSR budgets in more directions: 82.76% (donations and sponsorships), programs for employees (68.97%), community programs (41.38%), social causes (20.69%)
Corporate social responsibility awareness in Romania and Poland: A comparative analysis (Grabara et al., 2016)	A comparative study between Romania and Poland in terms of CSR initiatives and awareness among students regarding the topic. The authors highlight the role played by universities in educating younger generations and make them appreciate the important role of CSR for the community
Exploring the impact of corporate social responsibility policies on firm value: The case of listed companies in Romania (Gherghina & Vintila, 2016)	The authors notice the positive relationship between companies' CSR initiatives and their market value
Empirical study toward corporate social responsibility practices and company financial performance. Evidence for companies listed on the Bucharest stock exchange (Simionescu & Dumitrescu, 2018)	The authors analyze listed companies and find a positive correlation between the CSR programs and their financial results. The paper validates the hypotheses that show a direct relationship between specific types of CSR initiatives (focused on the environment, employees, clients, education, and healthcare) and the financial performance of companies
A human resources perspective on responsible corporate behavior. Case study: The multinational companies in western Romania (Obrad & Gherheș, 2018)	The study focuses on automotive multinational companies in the western part of Romania and especially on the internal dimension of CSR, which is oriented toward their own employees. The results show that companies in this sector understand the importance of taking care of their employees through dedicated CSR programs.

(continued)

Table 6.3 (continued)

Article	*Main ideas*
CSR development in post-communist economies: Employees' expectations regarding corporate socially responsible behavior—The case of Romania (Stoian & Zaharia, 2012)	The authors distinguish between endogenous and exogenous CSR from the perspective of the employees who work in multinational companies (more oriented toward CSR outside the company) and those who worked in socialist companies before 1989 (more oriented to an endogenous form of CSR)
A Country-Comparative Analysis of the Transposition of the EU Non-Financial Directive: An Institutional Approach (Aureli et al., 2020)	The authors compared Romania, Bulgaria, the United Kingdom, Belgium, and France in relation to the implementation of the non-financial reporting of CSR activities. The authors highlight the limited role of CSR in countries such as Romania and Bulgaria, which, from this perspective, are different from the other countries in the study that have intense CSR activity. Aureli et al. (2020) also notice the environmental initiatives of companies in Romania and the progress in this direction even if there are many challenges to be addressed, especially regarding the transparency of data regarding the impact on the environment.
Applications of TQM processes to increase the management performance of enterprises in the Romanian Renewable Energy Sector (Busu, 2019)	The author analyses the role of many factors in increasing the management performance of a company, and CSR is one of the factors alongside others such as "the integrated operational processes, trade strategies and policies, integrated operational management… motivated workforce, knowledge, and competencies" (Busu, 2019)
Non-financial information disclosures and environmental protection-evidence from Romania and Greece (Artene et al., 2020)	The authors show that the European Directive on non-financial reporting influences the listed companies to include more information regarding environmental protection in their sustainability reports and those in polluting sectors do this more than other companies
The impact of multinational companies from Romania on increasing the level of corporate social responsibility awareness (Dura & Driga, 2017)	The authors emphasize that multinationals in Romania have intense CSR activities while domestic companies do not fully understand the benefits of this type of involvement in the community. The authors notice that the awareness of CSR's importance increased and more companies dedicate budgets to CSR programs
Sustainability reporting is a mixture of CSR and sustainable development. A model for micro-enterprises within the Romanian forestry sector (Socoliuc et al., 2020)	The authors emphasize the importance of CSR for SMEs in the forestry sector in Romania and highlight the role the employees should play in integrating CSR into the organizational culture of these companies. The problem is important in the context of a sector that is fragile and faces many risks
Facilitators of and barriers to sustainable development in small and medium-sized enterprises: A descriptive exploratory study in Romania (Costache et al., 2021)	The authors analyze the factors that encourage CSR (the desire to do good, to leave something behind, to be proud of your work) but also those that act as an obstacle to these initiatives in SMEs in Romania (legislation, bureaucracy, funds)

(continued)

Table 6.3 (continued)

Article	*Main ideas*
Analysis of social responsibility and reporting methods of Romanian companies in the countries of the European Union (Socoliuc et al., 2018)	The study reveals the role of managers in Romanian companies in preparing CSR reporting. Those who have a proactive role are more inclined to report CSR activities in more detail in the annual reports, not perceiving this as just an obligation but as a way to promote the company
Demographics-based differences in the relationship between perceived CSR and customer loyalty in the dairy products market (Moisescu, 2015)	The author notices a positive relationship between customers' loyalty to dairy brands and the perception of the CSR activities of the companies in this sector. The relation is noticed when CSR is oriented toward the customers but for some categories of consumers, other types of CSR initiatives are also having a positive impact on their brand loyalty
Generating loyalty toward fast fashion stores: a cross-generational approach based on store attributes and socio-environmental responsibility (Dabija et al., 2022)	The authors emphasize the impact of CSR in the fast fashion sector on the loyalty of consumers. Consumers tend to be more loyal to stores that develop CSR initiatives to protect the environment or reduce their footprint
Can corporate fairness toward public authorities enhance customer loyalty? A multi-sectorial investigation in a developing country (Moisescu et al., 2019)	The paper finds a positive correlation between the perception of consumers regarding the good relationship between companies (from which they buy products and services) and public institutions and respectively their loyalty to these companies
The role of owner-managers in shaping CSR activity of Romanian SMEs (Saveanu et al., 2021)	The authors highlight the role of managers in the CSR strategy of Romanian SMEs. Those who understand the benefits of CSR and have the desire to give back to the community are also those with more CSR activities
The corporate social responsibility of the Romanian banking system (Dinu & Bunea, 2019)	The authors did not find a direct correlation between CSR reporting and the financial performance of banks in Romania but they emphasized the limits of their study (small sample and reduced timeframe)
The moderating influence of consumer demographics on the relationship between perceived CSR and brand loyalty in the Romanian retail banking sector (Moisescu & Gică, 2017)	The authors identified a positive correlation between the consumers' perception of the CSR engagement of banks in Romania and their loyalty to them
How well are manufacturing companies in Transylvania, Romania adapting to the low-carbon economy in order to become sustainable? (Dragomir et al., 2022)	The paper addressed the manufacturing sector in Transylvania and the readiness of these companies to change to a more low-carbon approach in the way they do business. These organizations understand the benefits of this type of economy but the financial resources needed to make the changes are also significant.

(continued)

Table 6.3 (continued)

Article	*Main ideas*
Sustainability and Social Responsibility of Romanian Sport Organizations (Barbu et al., 2022)	The authors highlight that CSR actions of sports organizations have a positive impact on their performance because of their visibility in the community
The relationship between corporate social responsibility and financial performance in Romanian companies (Matei et al., 2021)	The authors identified a positive relationship between compliance with the ISO 26000 standard and the financial performance of Romanian companies
OHS disclosures within non-financial reports: The Romanian case (Păun et al., 2020)	The authors show that managers in Romanian companies are becoming more aware of the role of their employees and, thus, they give more attention to the reporting regarding occupational health and safety
Corporate Social Responsibility in Liquid Times: The Case of Romania (Grigore et al., 2021)	The authors place CSR in Romania in a cultural and political context, taking into account the communist past of the country which changed the way people think, act, and live. The authors use the term liquid CSR to express the fact that CSR is more "promotional, bureaucratic, short-term, subject to arbitrary change, and outside of moral concerns" (Grigore et al., 2021)
Analysis of the influence of financial factors on corporate social responsibility: Evidence from Romania (Man et al., 2021)	The authors emphasize the impact of the financial performance of the listed companies as well as their size on the CSR activities they develop
Effect of pharmaceutical companies' corporate reputation on drug prescribing intents in Romania (Ion et al., 2021)	The paper analyses the relationship between the reputation of pharma companies and the intention of doctors to prescribe their medications. In this context, CSR is seen as a means to raise the reputation of a company and make it more visible and trustworthy

Among the factors that encourage managers to initiate CSR, the authors mention their desire to do good, the reputation and the image of the company in the community.

The second article (Socoliuc et al., 2018) refers to obstacles to CSR reporting and not CSR initiatives, so it is not so relevant for answering RQ2. But still, problems in reporting might influence the decision to develop CSR projects.

For answering RQ3, we searched the Scopus database and got five results after searching the keyword combination *CSR public sector Romania*. In the screening stage of the PRISMA method, we removed duplicates and the papers not meeting the criteria of time, language, and open-access status. We remained with only one result: *The moderating influence of consumer demographics on the relationship between perceived CSR and brand loyalty in the Romanian retail banking sector* (Moisescu & Gică, 2017). This is a duplicate of an article taken into account in the RQ1 analysis so it will be counted only once in Table 6.2, but we will consider it for the RQ2 analysis. After carefully reading the full paper, we understand that the authors analyzed the relationship between the CSR dimension of banks in Romania (including banks' responsibility to the public institutions) and the customers' loyalty toward them. So, the paper does not make any references to CSR in the public sector which is not so debated in the literature, being considered as more specific to the private sector.

For answering RQ4, we searched the Scopus database and got 15 results after searching the combination of keywords *CSR awareness Romania*. In the screening stage, we applied the exclusion criteria and got five articles: *The corporate social responsibility of the Romanian banking system* (Dinu & Bunea, 2019), *How Well Are Manufacturing Companies in Transylvania, Romania Adapting to the Low-Carbon Economy in Order to Become Sustainable* (Dragomir et al., 2022)? *OHS disclosures within non-financial reports: The Romanian case* (Păun et al., 2020), *The impact of multinational companies from Romania on increasing the level of corporate social responsibility awareness* (Dura & Driga, 2017) and *Corporate social responsibility awareness in Romania and Poland: A comparative analysis* (Grabara et al., 2016). All these articles are duplicates of the ones generated for RQ1 so they will be counted once but we will analyze them in order to answer RQ4.

Dragomir et al. (2022) address the problem of the carbon footprint of the manufacturing sector in Transylvania, their awareness of the way they affect the environment and their willingness to invest in more ecofriendly manufacturing solutions. The authors conclude that the awareness in this sector is high and managers are willing to share the knowledge with the employees but more actions are needed at the national level: better legislation, educational policies to raise the awareness of younger generations, adjusting universities' curricula to the nowadays realities of climate change and the need to move toward a low-carbon economy in general, not only in the manufacturing sector.

Păun et al. (2020) introduce the awareness concept in relation to the non-financial reporting of the CSR focused on the occupational health and safety of the employees. The authors notice the increased awareness of companies regarding the employees' working conditions and safety at work and the need to disclose these aspects in the annual CSR reports. Dinu and Bunea (2019) analyze the CSR of the banks in Romania through a complex CSR index with nine variables, each with equal shares. "Marketing and creating awareness" is one of the variables considered by the authors which highlights the actions taken by banks in this direction, of making people aware of the importance of CSR and thus improving their reputation in people's perception.

The other two papers have keyword *awareness* starting with the title. Grabara et al. (2016) made a comparison between Romania and Poland regarding CSR initiatives and conducted research to identify the level of awareness among students in a Romanian university. Their study highlights the need for introducing more courses related to CSR and sustainability in the academic curricula in order to raise CSR awareness and, indirectly, the CSR initiatives in the community. Two of the authors of the previous paper, Dura and Driga (2017) also analyzed the role of multinationals in Romania in raising awareness toward CSR aspects. Besides the multinationals, the authors emphasize the role played by public authorities in raising awareness. This is important in relation to RQ3 because the public authorities can facilitate

CSR initiatives through educational campaigns and improved legislation. Dura and Driga (2017) show that multinationals in Romania correlate CSR programs with an educational component to make younger generations more responsible and have a future-oriented perspective.

For answering RQ5, we searched the Scopus database using the keywords *CSR financial performance Romania* and got 10 results. In the screening stage, we applied the exclusion criteria and remained with six articles: *The relationship between corporate social responsibility and financial performance in Romanian companies* (Matei et al., 2021), *Sustainability reporting as a mixture of CSR and sustainable development. A model for microenterprises within the Romanian forestry sector* (Socoliuc et al., 2020), *An integrated framework on the sustainability of SMEs* (Burlea-Schiopoiu & Mihai, 2019), *The corporate social responsibility of the Romanian banking system* (Dinu & Bunea, 2019), *Empirical study toward corporate social responsibility practices and company financial performance. Evidence for companies listed on the Bucharest stock exchange* (Simionescu & Dumitrescu, 2018) and *Testing the correlations between corporate giving, performance, and company value* (Hategan & Curea-Pitorac, 2017). All articles are duplicates of those in Table 6.3 so we will count them only once in Table 6.2. All of them are recent papers that were published in the last 6 years (2017–2021).

All articles were read in their entirety to provide a better picture of the way the authors reflected in their papers the correlation between CSR and the financial performance of companies. Matei et al. (2021) analyzed the relationship between CSR and two financial indicators (return on assets and return on equity) and found a positive relation in both cases, this being a powerful motivating factor for companies to allocate more resources to CSR programs. Socoliuc et al. (2020) present other studies that found a positive relationship between CSR and financial results, but their study focuses on SMEs in the forestry sector and non-financial CSR reporting.

Burlea-Schiopoiu and Mihai (2019) focus on SMEs in the Oltenia region in Romania and the correlation between their CSR activities and financial results, emphasizing that the literature review usually studies big companies, not small and medium ones. The authors validated their hypothesis according to which CSR can help companies to have better financial results. Simionescu and Dumitrescu (2018) focus on listed companies and analyze the CSR strategy in relation to their financial performance. The authors concluded that companies with increased CSR activities also have better financial performance. Hategan and Curea-Pitorac (2017) conducted a similar analysis for listed companies and found a positive relationship between CSR and the return on equity.

All these studies show that the benefits of CSR for a company, no matter if small, medium or big, domestic or multinational, are not limited only to the improved reputation and image but extend to better financial performance too which is an important motivational factor for companies to be more socially responsible.

To answer RQ6, we searched the Scopus database using the keywords *CSR crisis Romania* and got eight results. In the screening stage, we applied the exclusion criteria and got two articles: *An integrated framework on the sustainability of SMEs* (Burlea-Schiopoiu & Mihai, 2019) and *The corporate social responsibility of the*

Romanian banking system (Dinu & Bunea, 2019). Both of them were counted only once in Table 6.2 because they are also duplicate articles which were found as results for the other questions. Burlea-Schiopoiu and Mihai (2019) do not focus specifically on CSR in times of crisis but mention other authors in the literature review section (Lin et al., 2019) to emphasize that "CSR activity is less vulnerable to economic shocks, financial crises or environmental disasters." Dinu and Bunea (2019, p. 121) focus on the Romanian banking sector, not especially on their CSR in times of crisis but analyze the literature review and highlight that CSR can be seen as the "most efficient instrument to help to improve the reputation" which is affected during financial crises, like the one in 2008.

Conclusions

The present chapter presents the results of the research regarding CSR in Romania considering the principles of the PRISMA method. Thus, we searched the Scopus database and found 161 papers addressing this topic. After screening the results, we excluded 133 papers that did not meet the criteria of language, open-access status (in order to be able to read them), and time (published after 2008). Thus, we read in their entirety 28 open-access articles which are detailed in Table 6.3. They approached the topics illustrated by all the six research questions we formulated (RQ1–6).

In conclusion, there is a certain level of awareness and both the government and the public sector make efforts to disseminate CSR knowledge but there are still things that can be done better (legislation, academic curricula, more initiatives, support for SMEs). We also notice that there is not a significant debate on CSR during crises and what happens to the budgets allocated to these initiatives, but by highlighting the positive connection between CSR and profits, maybe CSR can be the solution to tackle crises and other socio-economic challenges that the Romanian society faces.

The limitations of the present study refer to the fact that there is a possibility that we excluded documents that could have given a more complex answer to the questions addressed but because they were not open access, we were not able to read them and thus, we did not consider them in the final analysis.

Still, the results of the study are helpful in drawing a better picture of what has been written in this field and identifying the research gaps. These gaps can be used as a starting point by other researchers: CSR in the public sector and the analysis of the CSR budgets in times of crisis and the impact of CSR on the recovery of a company during and after an economic, social, health, or financial crisis. When we screened the results in the Scopus database, we had to exclude chapters written on the topic because they were not open access. This is a limitation of our study but also an important indicator of the importance of making research open access. This helps with the visibility of the research but also might help with disseminating knowledge and raising awareness on some crucial topics like CSR and sustainability.

References

Ali, F., & Sohail, M. (2018). Effects of corporate social responsibility on consumer purchase intention. *Pakistan Journal of Humanities and Social Sciences, 6*(4), 477–491. https://doi.org/10.52131/pjhss.2018.0604.0059

Artene, A., Bunget, O.-C., Dumitrescu, A.-C., Domil, A.-E., & Bogdan, O. (2020). Non-financial information disclosures and environmental protection—evidence from Romania and Greece. *Forests, 11*(8), 814. https://doi.org/10.3390/f11080814

Aureli, S., Salvatori, F., & Magnaghi, E. (2020). A country-comparative analysis of the transposition of the EU non-financial directive: An institutional approach. *Accounting, Economics, and Law: A Convivium, 10*(2). https://doi.org/10.1515/ael-2018-0047

Badulescu, A., Badulescu, D., Saveanu, T., & Hatos, R. (2018). The relationship between firm size and age, and its social responsibility actions—focus on a developing country (Romania). *Sustainability, 10*(3), 805. https://doi.org/10.3390/su10030805

Barbu, M. C. R., Popescu, M. C., Burcea, G. B., Costin, D.-E., Popa, M. G., Păsărin, L.-D., & Turcu, I. (2022). Sustainability and social responsibility of Romanian sport organizations. *Sustainability, 14*(2), 643. https://doi.org/10.3390/su14020643

Burlea-Schiopoiu, A., & Mihai, L. S. (2019). An integrated framework on the sustainability of SMEs. *Sustainability, 11*(21), 6026. https://doi.org/10.3390/su11216026

Busu, M. (2019). Applications of TQM processes to increase the management performance of enterprises in the Romanian renewable energy sector. *PRO, 7*(10), 685. https://doi.org/10.3390/pr7100685

Cambefort, M. (2020). How the COVID-19 pandemic is challenging consumption. *Markets, Globalization & Development Review, 5*(1). https://doi.org/10.23860/MGDR-2020-05-01-02

Cho, S. J., Chung, C. Y., & Young, J. (2019). Study on the relationship between CSR and financial performance. *Sustainability, 11*(2), 343. https://doi.org/10.3390/su11020343

Costache, C., Dumitrascu, D.-D., & Maniu, I. (2021). Facilitators of and barriers to sustainable development in small and medium-sized enterprises: A descriptive exploratory study in Romania. *Sustainability, 13*(6), 3213. https://doi.org/10.3390/su13063213

Dabija, D. C., Câmpian, V., Pop, A. R., & Băbuț, R. (2022). Generating loyalty towards fast fashion stores: A cross-generational approach based on store attributes and socio-environmental responsibility. *Oeconomia Copernicana, 13*(3), 891–934.

Danciu, V. (2021). Adapting and modernizing the post-Covid marketing policy under the impact of consumption challenges. *Romanian Economic Journal, 79*, 2–21. https://doi.org/10.24818/REJ/2021/79/01

Dinu, V., & Bunea, M. (2019). The corporate social responsibility of the Romanian banking system. *E a M: Ekonomie a Management, 22*(4), 119–133.

Dragomir, M., Blagu, D. A., Popescu, S., Fulea, M., & Neamțu, C. (2022). How well are manufacturing companies in Transylvania, Romania adapting to the low-carbon economy in order to become sustainable? *International Journal of Environmental Research and Public Health, 19*(4), 2118. https://doi.org/10.3390/ijerph19042118

Dura, C., & Driga, I. (2017). The impact of multinational companies from Romania on increasing the level of corporate social responsibility awareness. *Contemporary Economics, 11*(1), 45–67.

Gherghina, S. C., & Vintila, G. (2016). Exploring the impact of corporate social responsibility policies on firm value: The case of listed companies in Romania. *Economics & Sociology, 9*(1), 23–42.

Grabara, J., Dura, C., & Driga, I. (2016). Corporate social responsibility awareness in Romania and Poland: A comparative analysis. *Economics & Sociology, 9*(4), 344–359.

Grigore, G., Molesworth, M., Vontea, A., Basnawi, A. H., Celep, O., & Jesudoss, S. P. (2021). Corporate social responsibility in liquid times: The case of Romania. *Journal of Business Ethics, 174*(4), 763–782.

Gupta, S., Nawaz, N., Alfalah, A. A., Naveed, R. T., Muneer, S., & Ahmad, N. (2021). The relationship of CSR communication on social media with consumer purchase intention and brand admiration. *Journal of Theoretical and Applied Electronic Commerce Research, 16*(5), 1217–1230. https://doi.org/10.3390/jtaer16050068

Hategan, C.-D., & Curea-Pitorac, R.-I. (2017). Testing the correlations between corporate giving, performance and company value. *Sustainability, 9*(7), 1210. https://doi.org/10.3390/su9071210

Ion, L. M., Vodă, A. I., Butnaru, R. C., Butnaru, G. I., & Chirita, G. M. (2021). Effect of pharmaceutical companies' corporate reputation on drug prescribing intents in Romania. *Economic Research-Ekonomska Istraživanja, 34*(1), 521–544.

Lee, K. H., & Shin, D. (2010). Consumers' responses to CSR activities: The linkage between increased awareness and purchase intention. *Public Relations Review, 36*(2), 193–195. https://doi.org/10.1016/j.pubrev.2009.10.014

Li, J., Hallsworth, A. G., & Coca-Stefaniak, J. A. (2020). Changing grocery shopping behaviours among Chinese consumers at the outset of the COVID-19 outbreak. *Tijdschrift voor Economische en Sociale Geografie, 111*(3), 574–583. https://doi.org/10.1111/tesg.12420

Lin, L., Hung, P. H., Chou, D. W., & Lai, C. W. (2019). Financial performance and corporate social responsibility: Empirical evidence from Taiwan. *Asia Pacific Management Review, 24*(1), 61–71.

Man, M., Păunescu, C. F., & Bogeanu-Popa, M. M. (2021). Analysis of the influence of financial factors on corporate social responsibility: Evidence from Romania. *Polish Journal of Management Studies, 24*(1), 202–219.

Matei, F. B., Boboc, C., & Ghiţă, S. (2021). The relationship between corporate social responsibility and financial performance in Romanian companies. *Economic Computation & Economic Cybernetics Studies & Research, 55*(3), 297–314.

Mehta, S., Saxena, T., & Purohit, N. (2020). The new consumer behaviour paradigm amid COVID-19: Permanent or transient? *Journal of Health Management, 22*(2), 291–301. https://doi.org/10.1177/0972063420940834

Moisescu, O. (2015). Demographics-based differences in the relationship between perceived CSR and customer loyalty in the dairy products market. *Management & Marketing, 10*(2), 118–131. https://doi.org/10.1515/mmcks-2015-0010

Moisescu, O. I., & Gică, O. A. (2017). The moderating influence of consumer demographics on the relationship between perceived CSR and brand loyalty in the Romanian retail banking sector. *E a M: Ekonomie a Management, 20*(4), 187–202.

Moisescu, O. I., Gică, O. A., Müller, V. O., & Müller, C. A. (2019). Can corporate fairness towards public authorities enhance customer loyalty? A multi-sectorial investigation in a developing country. *Sustainability, 12*(1), 187. https://doi.org/10.3390/su12010187

Obrad, C., & Gherheş, V. (2018). A human resources perspective on responsible corporate behavior. Case study: The multinational companies in western Romania. *Sustainability, 10*(3), 726.

Page, M. J., McKenzie, J. E., Bossuyt, P. M., Boutron, I., Hoffmann, T. C., Mulrow, C. D., et al. (2021). The PRISMA 2020 statement: An updated guideline for reporting systematic reviews. *BMJ (Clinical research ed.), 372*, n71. https://doi.org/10.1136/bmj.n71

Păun, A. P., Dura, C. C., Mihăilescu, S., Moraru, R. I., & Isac, C. A. (2020). OHS disclosures within non-financial reports: The Romanian case. *Sustainability, 12*(5), 1963. https://doi.org/10.3390/su12051963

Puiu, S., & Wiśniewski, A. K. (2020). Corporate social responsibility of Romanian SMEs versus polish SMEs. In S. Idowu & C. Sitnikov (Eds.), *Essential issues in corporate social responsibility. CSR, sustainability, ethics & governance*. Springer. https://doi.org/10.1007/978-3-030-39229-1_3

Rhou, Y., Singal, M., & Koh, Y. (2016). CSR and financial performance: The role of CSR awareness in the restaurant industry. *International Journal of Hospitality Management, 57*, 30–39. https://doi.org/10.1016/j.ijhm.2016.05.007

Saveanu, T., Badulescu, D., Saveanu, S., Abrudan, M.-M., & Badulescu, A. (2021). The role of owner-managers in shaping CSR activity of Romanian SMEs. *Sustainability, 13*(20), 11546. https://doi.org/10.3390/su132011546

Sharma, V., Poulose, J., Mohanta, S., & Antony, L. E. (2018). Influence of the dimensions of CSR activities on consumer purchase intention. *Innovative Marketing, 14*(1), 23–32. https://doi.org/10.21511/im.14(1).2018.03

Simionescu, L. N., & Dumitrescu, D. (2018). Empirical study towards corporate social responsibility practices and company financial performance. Evidence for companies listed on the Bucharest stock exchange. *Sustainability, 10*(9), 3141.

Socoliuc, M., Grosu, V., Hlaciuc, E., & Stanciu, S. (2018). Analysis of social responsibility and reporting methods of Romanian companies in the countries of the European Union. *Sustainability, 10*(12), 4662.

Socoliuc, M., Cosmulese, C. G., Ciubotariu, M. S., Mihaila, S., Arion, I. D., & Grosu, V. (2020). Sustainability reporting as a mixture of CSR and sustainable development. A model for micro-enterprises within the Romanian forestry sector. *Sustainability, 12*(2), 603.

Stoian, C., & Zaharia, R. M. (2012). CSR development in post-communist economies: employees' expectations regarding corporate socially responsible behaviour–the case of Romania. *Business Ethics: A European Review, 21*(4), 380–401.

Ursic, D., & Cestar, A. S. (2022). Crisis management and CSR in Slovenian companies: The impact of the COVID-19 pandemic. *Sustainability, 14*(5), 2690. https://doi.org/10.3390/su14052690

Waworuntu, S. R., Wantah, M. D., & Rusmanto, T. (2014). CSR and financial performance analysis: Evidence from top ASEAN listed companies. *Procedia-Social and Behavioral Sciences, 164*, 493–500. https://doi.org/10.1016/j.sbspro.2014.11.107

Zhou, Z., & Ki, E. J. (2018). Exploring the role of CSR fit and the length of CSR involvement in routine business and corporate crises settings. *Public Relations Review, 44*(1), 75–83. https://doi.org/10.1016/j.pubrev.2017.11.004

Silvia PUIU, PhD, is Associate Professor, PhD Habil. at the Department of Management, Marketing and Business Administration within the Faculty of Economics and Business Administration, University of Craiova, Romania. She has a PhD in Management and she teaches Management, Marketing, Ethics Management in Business, and Creative Writing in Marketing. During the last few years, Silvia Puiu published more than 50 articles in national and international journals. Her research covers topics from corporate social responsibility, strategic management, ethics management, public marketing, and management. In 2023, she received the Outstanding Award for Excellence in Manuscript Reviews from the journal Sustainability which is indexed in the Web of Science database.

Chapter 7
Unpacking Corporate Social Responsibility Activities in Organisations by Considering Social, Economic, and Environmental Challenges: A Conceptual Approach

Afzal Sayed Munna, Md Sadeque Imam Shaikh, and Uzoechi Nwagbara

Introduction

CSR activities are seen to be performed by various organisations in order to promote better sustainability and cater support to the surrounding. Companies try to integrate different environmental and social concerns with their respective business operations and relevantly interact with various stakeholders to bring about important CSR activities (Rhee et al. 2021). This particular activity's benefit involves enhancing better public trust by bringing about an attractive perspective on behalf of employees. In addition to this, CSR activities also help in more engagement of employees and improve their overall morale by bringing about positive approaches. Different organisations seem to bring about various activities related to CSR as, to a large extent, it helps in improving the overall customer perception regarding the brand.

Four important types of CSR responsibilities can be observed in an organisation that leads to the betterment of the environment as well as for the organisation. Environmental, ethical, philanthropic, and economic responsibilities are relevantly involved within CSR that the organisation effectively follows to bring about necessary changes in the society. SMEs also seem to make a relevant contribution to the important actions done in the case of CSR. This study aims to focus on important CSR initiatives and methods used by various international organisations and SMEs of Bangladesh to bring about better development for the economy and environment. Within the context of this research, we specifically frame our literature analysis on

A. S. Munna (✉) · U. Nwagbara
University of Sunderland in London, London, UK
e-mail: Afzal.Munna@sunderland.ac.uk; uzoechi.nwagbara@sunderland.ac.uk

M. S. I. Shaikh
University of Wales Trinity Saint David, London, UK

U. Nwagbara et al. (eds.), *Corporate Social Responsibility Disclosure in Developing and Emerging Economies*, CSR, Sustainability, Ethics & Governance, https://doi.org/10.1007/978-3-031-61976-2_7

the textile sector for CSR initiatives since the industry plays a critical role in the development of the economic growth of Bangladesh (Rasel et al., 2020). The main reason for this claim comes from the economic nature of the country because it exports textile and garment products to different regions of the world. Based on the aforementioned sector perspective, this study considers CSR initiatives of various organisations related to the textile sector, including NGOs, to demonstrate multiple external challenges and key initiatives. Furthermore, the involvement of SMEs and the scope for further improvement of CSR activities are also elaborated on in this study. Therefore, the different sections of this particular study will provide a detailed understanding of the important activities in the textile sector necessarily do in order to provide better CSR initiatives and to help the business-oriented surrounding.

CSR in Bangladesh

Involvement of Corporate Code of Conduct

A Corporate Code of Conduct (CCC) is a standard followed by various organisations that aim to follow sustainable practices and bring about objectives related to the organisation's social responsibilities. As per the ideas of Joseph and Hashmi (2018), CCC is effectively followed by different organisations in order to integrate various market as well as non-market strategies. In the case of Bangladesh, different organisations presently follow this particular standard in order to maintain different kinds of activities related to CSR. This also provides a relevant way to maintain the strategic decision-making process within the organisation to create necessary changes within the business operations.

In most organisations, CCC mainly revolves around various financial strategies and purposes within Bangladesh. This is mainly because of the environment of financial activities within the organisation for conducting different types of CSR activities. From the viewpoints of Lee and Quan (2020), CCC is seen to revolve around various factors that influence CSR activities, which lead to financial gain within an organisation. This tends to be beneficial for an organisation in understanding important backdrops as well as legal regulations as those are essential for dealing with various activities related to social responsibilities. However, it can be relevantly estimated that the particular standard ranges from community involvement along with the inclusion of philanthropic donations, which brings profit to various organisations in Bangladesh. In addition, the standard involves bringing about excellent governance within the corporate environment, implementing better green strategies and policies, and finally bringing about environmental protection.

The inclusion of international trade and globalisation has instigated the relevant development of various interests in CSR as well as making significant growth in CCC. To a larger extent, this helps provide general information that involves companies and different small organisations more in CSR activities that initiate better development of Bangladesh. However, as identified in the findings of Velut et al. (2021), this standard is mainly concentrated on important economic sections of a country, such as the social and environmental aspects to a large extent. Therefore, this can be considered as one of the limitations of this particular standard that is often followed by various organisations often follow to bring about necessary activities of CSR. In order to improve CSR activities, organisations in Bangladesh need to initiate better changes in the environment. This can be relevantly done with the help of proper market research along with doing analysis of the important trends of CSR that can be observed in this particular country.

Role of MNC in Bangladesh

Different MNCs are seen to be present in Bangladesh and are seen to be following all the important activities that are related to the betterment of society and the environment. Based on the idea of Kim et al. (2018), the involvement of important activities seems to boost the economy of a country that an organisation often highlights. In the case of Bangladesh, it is seen that the inclusion of CSR activities involves bringing about better job opportunities for the youths, which are relevantly done by the MNCs present in the concerned economy. Organisations like Amul and Unilever that have their branches in Bangladesh are seen to bring about necessary improvement in different CSR activities to a larger extent.

Ethical responsibilities are effectively considered by the organisations present in Bangladesh to bring about development in society and stipulate better development of the economy. Identified in the findings of Siltaloppi et al. (2021), maintaining ethical standards of CSR and bringing forward better profits for the organisation are considered to be some of the important aims of CSR activities performed by any organisation. Various MNCs in Bangladesh are seen to develop different kinds of initiatives that enhance their overall brand reputation and bring about better profit in the competitive market. For improving the overall organisation activities and bringing financial gain, different kinds of social and environmental responsibilities are being followed in order to enhance the overall growth of the company.

Various MNCs that are involved and seem to be working efficiently in Bangladesh are seen to highlight their respective commitment towards sustainability and good CSR practices. According to the ideas of Perry and Wood (2019), organisations also focus on non-financial activities and responsibilities that are related to society or the environment. A proper voluntary contribution is seen to be made by various initiatives in order to enhance better CSR activities within the economy of Bangladesh. On the contrary, Bangladesh often faces legitimacy issues because of the political and socio-environmental conditions of the economy. This is major because of the uncertainty and discontinuous functions that are often observed within the concerned country by bringing about legal pursuits to a large extent. Therefore, it can be relevantly said that the involvement of different MNCs in Bangladesh can bring forward necessary CSR activities that help boost the economic development of

Bangladesh by considering non-financial areas such as human rights, business ethics, and environmental policies that come from the integration strategies to the world market (Alam et al., 2010).

Different Regulations of CSR Present in Bangladesh

Regulations are the same that initiate various activities related to CSR in Bangladesh and are considered to be essential as it initiates better activities by the organisation. According to the ideas of Beckers (2019), the environment of corporate self-regulation and internal strategies act as a significant regulation for bringing about different kinds of CSR activities in a particular country. In the case of Bangladesh, it is observed that a proper corporate culture is essential for bringing about corporate self-regulation that enhances various activities, meeting the needs of the environment and society in a relevant way.

In Bangladesh, it is observed that fewer specific regulations or laws are present related to CSR activities that integrally identify different issues present within the strategies. Three important laws are involved in Bangladesh that tend to advocate and promote CSR activities which are *"Companies Act 1994"*, *"the Bangladesh Labour Act 2006"*, and finally *"the Environment Conservation Act 1995"* (Beza.gov.bd, 2022). All these are considered to be some of the important laws that are relevantly followed by various organisations present within Bangladesh in order to approve various CSR activities. Issues related to CSR activities have also been identified with this standard and a proper aggregation of the activities that are often being made by the government present within the country.

Different organisations in the textile industry operating in Bangladesh are seen to have a detailed prospectus where information related to CSR activity is mentioned along with the strategies for the further improvement process. The mention of different information and activities within an organisation to the general public helps in boosting the reputation of an organisation and creates a positive approach towards customer perception (Zuraida and Sugianto, 2021). The environment of different directors within the corporate regulations is effectively mentioned in prospectors by bringing about an ability to understand the important compensations made by investors in case of any mishap. A proper focus is given on the labours as most of the activities are done, keeping in mind providing equal opportunity to the employees present within the organisation along with focusing on boosting employment. Hence, the involvement of important regulations and their CSR brings about a proper structure to the organisations of Bangladesh to deal with important activities and to show a positive responsibility towards the society.

External Challenges Faced by Bangladesh

Social Challenges

Some challenges in the social aspects can be relevantly observed by various enterprises present in Bangladesh as it is a developing country and requires maximum revenue generation in order to mitigate the challenges. Social welfare involves setting up a better work-life balance, which is considered one of the important challenges in Bangladesh's economy regarding the proper implementation of CSR activities. This is mainly because of the less revenue that is often generated from different organisations and more work pressure often given to the employees to complete the pending works. Based on the ideas of Lu et al. (2019), the working environment plays a crucial role in dealing with various CSR activities within an organisation that motivates employees to work in a positive mindset. For this reason, a bad working environment is mainly a result of poor investment that further acts as a significant challenge for organisations to react to different activities positively. In addition to this, lesser information about the wage gap is also involved within the organisations of Bangladesh, which seems to bring about a challenging situation in understanding the needs of the employees. Bangladesh has occupational wages and the nature of employment that often creates a hectic situation for the organisation to segregate employees for their skills and knowledge. This creates a negative attitude towards the organisation that affects the overall CSR activities concerned by any organisation.

Economic Challenges

Different developing countries, such as Indonesia and Malaysia, are seen to have relevant economic issues for which they face challenges in the proper implementation process of CSR. As per the ideas of Indriastuti and Chariri (2021), a proper investment within organisations for different business operations helps bring about innovative and creative business operations helps in improving profit and supporting the environment. In the case of Bangladesh, lesser investments by various organisations are seen to be banned by the investors due to a lack of proper financial aid. This creates a relevant challenge for the organisations to bring forward various CSR activities that are going to be beneficial for the organisation as well. In addition to this, a lack of proper awareness amongst the different stakeholders also contributes to this economic challenge faced by various businesses for CSR. The recent pandemic situation has also created a relevant challenge to the different organisations in bringing forward important strategies related to CSR because of the lack of proper investment done by various stakeholders to a larger extent. A better analysis of the entire scenario is essential to deal with the economic challenge faced in bringing CSR activities and helping the country (Amor-Esteban et al. 2020). Organisations in Bangladesh are trying to analyse important economic issues by bringing about solutions like foreign direct investments and conducting awareness programs for the stakeholders in order to enhance the overall investment for CSR activities. However, the government also plays a significant role in directly connecting with various CSR activities, which is further seen in Bangladesh as some of them are partners with the government and respective organisations.

Environmental Challenges

Corporate social responsibility plays a vital role in eradicating environmental issues in the present context of increased pollution. Rapid urbanisation and industrialisation have generated major environmental problems within Bangladesh, such as soil, air and water pollution, ineffective waste management, and noise pollution. As per the ideas of Mitu et al. (2022), the country is also facing radical environmental issues due to groundwater metal contamination, changing patterns of streamflow, increased groundwater salinity, and watershed mismanagement. These attributes placed the country at a high risk of global warming and climate change. In addition, increased population, associated motorisation, and industrialisation have caused massive air pollution, based on which the country is facing a severe challenge of maintaining the air quality index. The country's capital, Dhaka, has reached its yearly average of 90 $\mu g/m^3$ of PM 2.5, which can be considered an unhealthy air quality (Aqicn.org, 2022). Due to these attributes, the country can impose measures regarding energy consumption control and waste management efforts on SMEs, which direct the economy towards instability. However, these challenges are also impacting the financial flow of organisations by deriving additional expenses through the development of waste management and procuring renewable energy utilising devices.

Environmental adversities such as depleting water levels and ineffective disposal of toxic as well as chemical waste are impacting the credibility of organisations in performing CSR. Based on Scherer and Voegtlin's (2020) views, business firms are accountable for societal and environmental change due to their business operations. The awareness of environmental sustainability is also one of the primordial challenges based on which SMEs and regional businesses can face issues regarding business operations. Since Bangladesh is the sixth most natural disaster-prone country across the globe, diverse meteorological and hydrological states can be seen within the country (Rahman et al., 2019). It has been found that an average of five severe cyclonic storms hit Bangladesh every year, which has impacted the business operations negatively (Rahman et al., 2019). Severe flooding has also been seen within several plateau regions due to huge astronomical tides in combination with cyclone surges. Though the government has not imposed any explicit policy on CSR for SMEs, national banks of the regime have mandated CSR for the firms in order to obtain business insurance. However, the Ministry of Commerce and Information is encouraging private organisations to adopt CSR activities in order to contribute to the development of a robust mechanism for handling the losses caused by environmental hazards.

Political Challenges

Political instability and disruptive governance practices can impact the credentials towards performing CSR within an organisation due to lack of financial orientation and uncertainties. As per the views of Alam (2018), the challenge of fulfilling the needs of democratic governance is the primary issue in Bangladesh due to unstable political calibrations. Since political pluralism as well as participation has become limited, challenges towards obtaining political benefits and self-regulations towards maintaining the financial flow of enterprises can be impacted drastically. Lack of democratic practices and rising military coups have deteriorated the administration's power towards providing strategic direction for performing CSR.

Based on the ideas of Hoque (2018), it has been found that social discrimination is a part of political practices in Bangladesh, based on which a male supremacy can be seen. Gender equality has become a political issue in Bangladesh due to the rising practices of dowry-related violence, harassing women, and child marriage. It has been found that Bangladesh has one of the highest child marriage rates in the world, based on which issues such as improper education and unemployment can be seen (Plan-international.org, 2022). However, these challenges are impacting the ability of the government to analyse the requirement of CSR within industries, based on which it can bring sustainability.

Political CSR involves addressing regulatory gaps within the administrations caused by insufficient social or environmental standards as well as norms. The challenge of obtaining a transparent form of political lobbying has been faced by Bangladesh, which has impacted the social urgency towards adopting CSR within business initiatives. For instance, petrochemical organisations such as "*Super Petrochemical Ltd.*" and construction companies such as "*Bangladesh Engineering and Construction*" have eliminated CSR activities from business operations in order to maintain financial flow and political insurgency within the state (Sppl.com.bd, 2022). However, these political constraints can be eliminated by obtaining regulatory policies on CSR adoption, based on which SMEs can understand the importance of performing social responsibilities in order to bring sustainability. For this reason, political challenges can be a guide for politics to create regulations to remove barriers to CSR activities because such challenges prevent management boards in companies to decide some certain CSR practices such as raising issues related to gender diversity and minorities.

Key Initiatives for CSR and Sustainable Development

Philanthropic Responsibility

Philanthropic responsibility includes funding education initiatives, providing support to community beautification projects and supporting health initiatives. As per the ideas of Glonti et al. (2020), increasing unemployment and poverty in Bangladesh require philanthropic responsibility management in order to fulfil the regulatory gaps within societal conjectures. For instance, "*Navana property group*", which is a construction firm in Bangladesh, has pledged to contribute towards social causes as per their annual profit margins in order to increase the adoption of CSR within the particular industry (Navanapropertygroup.co.uk, 2022). The firm has also incorporated several CSR regulations of UNESCO such as workplace equality maintenance and waste monitoring in order to maintain ethical; gestures of CSR initiatives. However, the organisation has donated an estimated €22,000 to a kidney foundation in Bangladesh in order to promote CSR within the regime (Navanapropertygroup.co.uk, 2022). However, these initiatives can bring financial support as well as sustainability within the country, based on which effective measures towards CSR value acquisition can be obtained within major industries in Bangladesh.

Environmental Responsibility

In the present context of industrialisation, climate change and global warming are the primordial concerns that can bring operational constraints. Since Bangladesh is a natural disaster-prone country, a major portion of the financial budget is spent by the government on disaster planning and fund relief. For instance, *Bashundhara Group*, a Bangladesh-based property development organisation, has constructed the largest corona treatment hospital in Bangladesh (Bashundharagroup.com, 2022). The organisation has also developed regulations of CSR towards providing armed forces, law enforcers as well as coast guards with PPE during the pandemic. This particular initiative has influenced several SMEs in Bangladesh such as *Beximo and City Group Ltd.* to contribute to a national as well as global cause.

Waste control management can bring sustainability within countries, based on which governments can provide food to the citizens. As opined by Mohiuddin et al. (2021), due to cyclonic storms, floods, and heavy rains, Bangladesh used to suffer from food scarcity, which impacts the process of safeguarding the citizens from food uncertainties. On the other hand, the Rohingya crisis has infuriated the food supply chain of the country, based on which the government is spending €1.3 billion every year (Thediplomat.com, 2021). Language and religious sentiment have led the government towards providing shelter to the refugees of Myanmar, which has also been supported by several business firms in Bangladesh such as "*Magnito Digital Ltd. and NewsCred*" (Thediplomat.com, 2021). However, these activities promote CSR in Bangladesh, based on which an effective CSR mandate can be generalised by the administration for obtaining sustainability.

Workers Wellbeing Responsibility

Responsibility towards managing the wellbeing attributes of labourers and business professionals is also a primordial aspect of sustainability, based on which organisations can obtain effective reputational growth. For instance, *Datasoft Systems BD Ltd.* has initiated a worker wellbeing initiative in order to fulfil the requirements of the workforce to bring workplace sustainability (Datasoft-bd.com, 2022). Obtaining feedback from the employees regarding job responsibility drawbacks, social and official conjectures and physical as well as mental conditions can enhance the productivity of organisations as a whole. The firms can also incorporate information regarding financial and social requirements through deriving wellbeing approaches, which can create sustainable stakeholder relationships and eliminate workplace conflicts within the industries of Bangladesh. However, developing effective communication mediums can help the firms of Bangladesh to incorporate wellbeing attributed of employees, which can streamline the workforce with CSR.

On the other hand, a *holistic wellness strategy* is a derivational as well as conventional method of increasing employee participation. Firms across the globe utilise this philosophy within the workplace in order to reduce employee turnover and increase job satisfaction. This particular approach can indulge employees in volunteer participation in CSR activities of the firm irrespective of systematic conjectures. As per the ideas of Su and Swanson (2019), providing health benefits such as health insurance and fitness facilities to employees can result in increased job satisfaction, which can bring further possibilities for talent acquisition. It is important to consider the emotional metrics of employees in order to fulfil their abstract requirements during performing the tasks based on which productivity can be increased. However, new talents can be acquainted by firms through incorporating employee wellness responsibilities in the present context of business.

Ethical Responsibility

Ethical responsibilities deal with ensuring businesses towards obtaining fair trade practices across the board. This responsibility focuses on treating employees, stakeholders as well as consumers ethically, based on which a robust business mechanism can be developed. For instance, purchasing fair trade products and utilising renewable energy sources and equipment can be obtained by organisations in the textile industry to maintain social responsibilities in the present context of business. Due to a lack of administrative conjectures, SMEs of Bangladesh have been directed towards cheap management establishments, based on which ethical measures have been eliminated from the industries (Pittaway and Montazemi, 2020). Organisations can incorporate ethical responsibilities of recognising and interpreting multiple principles of work breakdown measure establishment and social conjectures in order to gain grounds within the present approach towards CSR development. However, rewarding ethical behaviour and providing training on ethical measure development can enhance the capability of organisations towards acquiring ethical responsibility.

Methods Used by Different International Organisation and NGOs to Boost CSR Initiatives

Methods Used by International Organisations

The banking sector of Bangladesh has an astonishing history of getting involved into CSR activities through their social practices such as state beautification, patronising cultural events, and providing financial assistance towards diverse groups. *Agrani Bank Ltd, Janata Bank Ltd,* and *Bangladesh Development Bank* have been advised by central administrative guidelines to be involved in these events in a more organised way to promote sustainability. In 2014, the *Sonali Bank Ltd* of Bangladesh contributed an estimated €452,000 towards the development of every field such as education and research, combat against natural calamity, and expansion of technology (Sonalibank.com.bd, 2021). This contribution has increased voluntary participation in the banking services, which creates a social conjecture towards performing CSR to support the economy of the nation.

It is important to focus on the development of a range of sectors within a regime to bring sustainability and influence the population's lifestyle. It has been found that Janata Bank Ltd. has donated an estimated €1.28 million to bring development within the health measurement as well as preservation of the environment sector (Janatabankpune.com, 2020). CSR activities related to poverty reduction and providing rehabilitation have also been incorporated by the banking sector of Bangladesh, based on which effective change within the CSR manifestation can be seen. As per the ideas of Anser et al. (2018), it is important to consider invention and technological advancement within the focus of CSR in order to bring a robust growth towards sustainability in the present context of business. However, the corporate sector of Bangladesh has also initiated several programs in order to provide employment to people, which has brought a social conjecture of self-governance in the present context of globalisation.

Rupali Bank Ltd. has also initiated several fundraising programs and CSR schemes towards their clients to uplift the capability of the government towards maintaining the development procedures of several sectors such as natural disaster management and income generation. Rupali Bank has donated €233,000 to several social causes such as poverty management and income generation schemes, which impacted the perceptions of other sectors towards CSR maintenance (Rupalibank.org, 2020). The bank has initiated a fundraising program to promote the classical music of the country, which helped more than 55,000 people find roots in classical music (Rupalibank.org, 2020). On the other hand, Citi Bank of Bangladesh has also contributed €44,600 in order to develop professionally qualified as well as competent bankers through training, examining, and development programs (Citigroup.com, 2022). This particular initiative has raised the practice of government competitive examinations in Bangladesh, through which talent can be trained effectively and by directing towards sustainable career options.

Method Used by NGOs

Textile associations such as the Bangladesh Garment Manufacturers and Exporters Association (BGMEA) and Bangladesh Knitwear Manufacturer and Exporters Association (BKMEA) are associated with philanthropic as well as environmental CSR activities. These two firms have initiated a scholarship distribution program in Bangladesh called Babylon Scholarship in order to promote basic education to the citizens (Zanalaarchive.com, 2022). The primordial aspect of this particular initiative is to acknowledge underprivileged talented students of the country to provide the facility of higher education. An estimated 90,000 students have been provided with this scholarship as of 2020, which has influenced the CSR approach of SMEs and other unstructured businesses. The joint venture has developed a budget of €1.3 million in 2020 for performing scheduled CSR activities (Zanalaarchive.com, 2022). However, these initiatives can bring a sense of contribution within other industries, based on which environmental as well as economic sustainability can be obtained.

Method Used by NGOs

Getting involved with local communities, volunteering, and fundraising are the primordial methods of performing CSR, based on which effective change can be obtained. Alternative transport systems and employee satisfaction are also part of CSR activities that can be performed by organisations effectively. NGOs have the ability to perform CSR selflessly due to their attribute towards volunteering. For instance, *Caritas Bangladesh* is an NGO which has initiated several educational schemes with the government in order to provide education towards underprivileged children (Caritasbd.org, 2022). The organisation has adopted the process of acknowledging underprivileged children through school surveys, based on which hidden talents can be obtained for this particular scholarship program. However, these attributes can increase the involvement of several industries in performing social responsibilities to promote development within the country.

Due to the ineffective legal framework and small government capacity, NGOs in Bangladesh play an essential role in addressing legal as well as political issues such as current government policies and economic issues. Services provided by NGOs of Bangladesh such as BURO and Oxfam are associated with economic development, employment, and budget stimulation (Oxfam.org, 2022). These NGOs have also promoted significant environmental problems in order to assist the disaster relief initiatives of the government during environmental emergencies such as floods and cyclonic storms. For instance, BRAC Bangladesh is one of the leading NGOs, which is associated with a diverse range of social issues such as poverty reduction and providing education and health services to the needy (Brac.net, 2022). The organisation provides microloans with zero interest to uplift the population's self-confidence, based on which effective change within the regime can be seen. The NGO has developed a CSR model to show how the competitive advantages of NGOs can be functioned towards supporting poverty reduction initiatives and human rights services (Brac.net, 2022). However, these NGOs have also been associated with health drives across the country such as Malaria and Dengue campaigns in order to provide fundamental medicine and food supply to the population during emergencies.

The role of NGOs can be divided into three main sectors such as service delivery, catalyst, and partnership-based approach. These predominant roles can be associated with a large number of participation, which can enhance the profitability process within CSR initiatives. For instance, activities of service delivery by NGOs built trust within local communities, based on which regional platforms can be created for community organising and advocacy. *BRAC and ASA* have developed combined approaches towards providing fundamental education to the children of Bangladesh, in which three million children have been enrolled, and more than 1000 libraries have been established (Brac.net, 2022). BRAC has also initiated Oral Therapy Extension Programs (OTEP) in order to reduce infant mortality rates in Bangladesh. Microloans provided by the NGO have helped SMEs and unstructured businesses in Bangladesh and have influenced several industries to perform CSR activities for the development of the country (Brac.net, 2022). NGOs such as *Shakti Foundation and Oxfam* have also responded to several damages caused by cyclonic storms in order to provide the residents with basic support (Shakti.org.bd, 2022). However, these NGOs have also responded to global warming and air pollution issues in Bangladesh, which can bring sustainability within the process of rapid industrialisation.

SMEs Attempt to Promote CSR Initiatives in Bangladesh

CSR is relatively a new subject within the unstructured industries in Bangladesh, which has directed the concept towards economic conjectures only. In Bangladesh, state-owned enterprises and commercial banks believe in CSR capabilities, based on which the firms have obtained regular annual development for the development of the country. Bangladesh has an estimated 7,900,000 SMEs, of which 93.4% are small, and 6.4% are medium businesses (Plandiv.portal.gov.bd, 2022). SMEs of a range of sectors such as website development, fabric stores, and digital studios are coming up with CSR initiatives jointly in order to bring sustainability in the present context of business. As per the ideas of Sobhani et al. (2021), consumer services groups such as *Agamee Prakashani, Adamjee Jute Mills, and Aarong* have incorporated a CSR budget plan within their annual budget development in order to promote philanthropic CSR. On the other hand, conglomerates such as *Akij, Nanda Group, and Advanced Chemical Industries (ACI)* have also initiated several community programs in order to bring talents within the workforce to promote CSR (Aci-bd.com, 2022). However, these initiatives have influenced several companies to take part in CSR activities in order to bring sustainability to Bangladesh through development.

Organisational benefits such as reputation, trust, motivation, and cost reduction in business operations can be obtained by SMEs through performing CSR. Small and medium-sized businesses can obtain a brand recognition by making contributions towards social causes such as global warming, climate change, and unemployment. For instance, *Ananda Shipyard and Shipways*, a Bangladesh-based commercial vehicle and truck manufacturer, has pledged to make a CSR budget annually to compensate for the air pollution done by the firm (Anandashipyard.com. 2022). Asian TV is also Bangladesh-based broadcasting and entertainment organisation, which has contributed towards eradicating unemployment within the country by initiating employment drives every year (Asiantv.com.bd, 2022). These initiatives have influenced other SMEs to take part in performing CSR, based on which social as well as environmental development can be obtained within the country. However, constant donations have also been seen by the commercial banks of Bangladesh, which influenced the NBFCs to take part in CSR activities for uplifting the economy of the country.

CSR can ensure the development of quality products within SMEs, thereby improving the overall environment and social surroundings to provide excellent customer service. As per Manzoor et al. (2019), CSR plays a vital role in developing an ethical understanding of the impact of large business operations on the environment within SMEs, based on which businesses can derive environmentally friendly operations. For instance, the textile industry of Bangladesh has initiated a socially conscious and ecofriendly clothing development program to reduce the carbon footprint caused by dying and synthesising the fabrics. Al Masud and Islam (2018) opined that companies such as *EOS Textile, Simco Spinning and Textile, and Salek Textile Ltd* had supported the initiative, bringing environmental sustainability to the regime. Thereby, it is important to analyse the carbon emission of manufacturing companies to depict a clear understanding of environmental pollution levels. An estimated 0.96 tons of CO_2 emission per capita has been recorded in Bangladesh due to the increased rate of industrialisation and decreasing social responsibilities (Climate-transparency.org, 2022). However, emerging trends of entrepreneurial business development focus on CSR incorporation, which can direct industries of Bangladesh towards obtaining sustainability and brand reputation.

On the other hand, it has been found that CSR activities can provide better financial performance, increased capability of attracting talents and organisational growth. Companies from the petrochemical industry of Bangladesh such as "*Super Petrochemical Ltd, Bangladesh Petrochemical Company Ltd, Chemtrek Industries Bangladesh Limited* and *CVO Petrochemical Refinery Ltd*" have initiated relief measures during flood and cyclonic attacks in order to safeguard the life of the population (Supremepetrochem.com, 2022). "*Super Petrochemical Ltd*" has also been associated with the construction of sanitation facilities, public waiting rooms in bus stands, and study halls for students in schools, which has increased the brand reputation of the firm within the petrochemical industry of Bangladesh (Supremepetrochem.com, 2022). These initiatives have influenced other petrochemical organisations such as *Matex and Rhymar Chemicals* in Bangladesh to incorporate CSR within business operations in order to bring social as well as environmental development within the country. However, developing risk management programs and community connections can enhance the potential of small businesses towards performing CSR in the present context of economic stimulation in Bangladesh.

Attributes such as increasing transparency, focusing on equity, and improving responsiveness can direct organisations towards CSR, based on which effective business management can be obtained. Based on the ideas of Radovanović et al. (2022), organisations can indulge their employees in CSR activities in order to bring ethical understanding as well as cultural diversity within the workforces. This particular initiative can provide self-satisfaction towards employees by providing opportunities towards contributing to sustainable developments. For instance, *Bashundhara Group* has initiated a health drive during the COVID-19 pandemic through employee volunteership in order to provide isolation as well as basic medication to patients (Daily-sun.com, 2022). On the other hand, *Beximco Pharmaceuticals* has also initiated several flood relief camps during cyclonic attacks on the country through engaging the workforce within the CSR programs (Beximcopharma.com, 2022). These initiatives have influenced other SMEs within the regime to obtain employee satisfaction and volunteer ship through CSR programs. However, indulging the workforce within a teamwork-based program can enhance the communication and team working ability of the workforce, which can bring sustainability within organisations.

SMEs implement CSR initiatives through several attributes such as getting involved with local communities and providing alternative transport methods. As per Graafland's (2020) ideas, small businesses demonstrate CSR commitments by developing local communities' conjectures such as education, employment, and food supply. It has been found that SMEs mostly work with local NGOs in order to articulate the requirements of the area, based on which sustainable artefacts related to mass development can be obtained. SMEs of Bangladesh also focus on the development of employees in order to bring sustainable production measures during harsh deadlines. Since a large population of Bangladesh is associated with daily labourer ship, indigenous as well as specific coordination is the primordial objective of SMEs in Bangladesh to maintain work ethics. Offshore and onshore productions can impact the CSR budget of SMEs due to limited risk management capabilities in the present context of business (Graafland, 2020). However, measures of developing healthy relationships amongst stakeholders and outsourcing partners can also be seen within SMEs of Bangladesh, based on which effective change can be obtained.

Due to the rapid change in labour costs, firms have outsourced their businesses within underdeveloped countries such as Bangladesh. As per Zatwarnicka-Madura et al. (2019), this particular initiative has impacted the attribute of mass unemployment (). Deterioration of daily wages and increased inflation have brought severe challenges to incorporating CSR within organisations, based on which inadequate measures towards social development have been seen. However, it is important to incorporate social concerns such as unemployment and health irregularities due to viral diseases such as Malaria as well as Dengue to bring a robust understanding of CSR approaches within SMEs and MNCs of Bangladesh (Zatwarnicka-Madura, 2019). Besides this, SMEs can obtain joint ventures with national and global NGOs to contribute to the development and bring brand reputation within the business. Following the discussion on promoting CSR activities in Bangladesh by SMEs, the next section includes further discussion on improving CSR activities through considering the scope of businesses.

Further Scope for Improvement of CSR

Corporate social responsibility has evolved into a medium of corporate life and has also become a part of corporate performance review. At the most basic level, CSR involves companies that want to make changes within business operations to obtain profit by not harming people as well as the environment. As per Kwarcinska's (2021) views, CSR can provide the scope of managing environmental aspects such as climate change, global warming, and regional calamities, based on which natural resources can be safeguarded. On the other hand, companies can promote renewable energy instead of fossil fuel-based energy consumption, which can increase the environment's capacity towards being more reproductive. Furthermore, based on the views of Kwarcinska (2021), cope of fair business practices such as employment and advancement of minorities as well as women can be performed by SMEs of Bangladesh to maintain ethical responsibilities. However, CSR initiatives can also bring ways to manage human resources more efficiently through wellbeing practices within the workforce.

On the other hand, CSR activities concerned with human resource management, such as recruiting practices, training programs, and job enrichment can establish greater accountability for being responsive within workforces. As opined by Gorgenyi-Hegyes et al. (2021), attributes such as wages and salary levels, fringe benefits plans and job security can be maintained by SMEs as well as MNCs through performing CSR activities. Practising volunteer ship within CSR programs can encourage employees to enhance their team working abilities effectively. CSR initiatives related to employee wellbeing can bring sustainable transformation as well as promotion policies within the workforce, based on which effective change can be obtained (Gorgenyi-Hegyes, 2021). Since CSR is comparatively a new subject in Bangladesh, it requires mass promotion, through which the effectiveness of CSR can be spread within a range of industries. CSR initiatives can allow the Bangladesh government to increase the public policy capacity and by freeing up existing resources, based on which robust economic growth can be seen.

CSR is associated with increasing social value and improved social welfare that can constitute the primordial objectives of government policies. As per the ideas of Rahman and Islam (2019), collaborative schemes can be created through CSR approaches with Bangladesh in order to provide basic supplies such as food as well as medicines to the people during natural calamities and national emergencies. On the other hand, philanthropic CSR initiatives can reduce costs as well as waste, which can enhance the economic capacities of countries. Since rapid industrialisation is affecting the environmental attributes of Bangladesh, a transverse waste disposal system is required in order to safeguard the agriculture sector of the regime. Natural disasters and economic depressions can be eradicated by developing collaborative CSR approaches, which can be used as special aids towards monitoring social issues (Rahman and Islam, 2019). Since most of the company focuses on employee loyalty and goodwill, economic attributes such as employee turnover and increasing productivity can be maintained effectively. However, promoting employee goodwill and attracting investors can also be a further scope of performing CSR, based on which effective change within the economic infrastructure can be obtained.

Community development is also a major scope of CSR activities, based on which requirements of the backward population of a country can be promoted. Activities such as obtaining education as well as health derived for tribal communities can be an effective measure towards community development, which showcases the ability to perform ethical responsibilities of firms in the present context of social biases. Since Bangladesh has one of the highest rates of child marriage. Promoting education and health requirements can eradicate the practice of child marriage from the regime (Sultana, 2021). Organisations can focus on consumer service management, based on which the quality of products, as well as services, can also be improved. As per the views of Roy and Quazi (2021), grants from the administration regarding product quality improvement can also be provided to the industries in order to enhance their performance towards obtaining an infuriated CSR budget. However, activities pertaining to social welfare as well as charitable initiatives can eradicate extreme hunger and poverty in Bangladesh and can also promote gender equality, which can bring sustainable measures within the regime in an effective manner.

Conclusion

Henceforth, it can be concluded that CSR initiatives are one of the leading edge mechanisms of uplifting performance, brand recognition as well as the economy within industries in the present context of business. Though the concept of CSR is relatively new in Bangladesh, SMEs and MNCs have penetrated several areas of social development, such as unemployment, health care, and education. It has been found that banking corporations of Bangladesh have been the major contributors towards CSR, which directed several business clients of the banks towards incorporating the responsibilities as well. In this particular study, different CSR practices of Bangladesh SMEs as well as MNCs have been discussed in order to establish a robust outline of the environmental and economic performance of the country. External challenges associated with social, economic, environmental as well as political issues have also been highlighted in this study to create a conceptual understanding of CSR management in Bangladesh. However, key initiatives of multinational companies and SMEs towards CSR and sustainable development have also been discussed in this study in order to understand the business environment of Bangladesh.

In the present context of business, CSR is more than just an overall business trend or fad, based on which innovative pathways for tackling environmental and social issues can be generated. It is an approach towards deriving sustainability for future generations, based on which civilians can obtain effective life measures. Initiatives towards the country's overall development in the field of innovation, education, and employment have been discussed in this study to create a robust understanding of CSR embracement. Methods utilised by different international organisations and NGOs for boosting CSR initiatives within the industries of Bangladesh have also been discussed in this study to outline the drawbacks of the initiatives. In addition, attempts of several SMEs towards promoting CSR initiatives have also been addressed in this study to bring a clear direction towards incorporating social measures. However, future scope associated with CSR has also been discussed in this study to enhance pathways of managing business operations ethically within the industries of Bangladesh.

References

Aci-bd.com. (2022). *Annual report 2020–2021*. Accessed May 20, 2022, from https://www.aci-bd.com/assets/files/financial/annual-report/aci_annual_report_2020_2021.pdf

Al Masud, A., & Islam, S. (2018). *The relationship between corporate social responsibility (CSR) and financial performance of textiles sector in Bangladesh especially in Barisal division*. IJASSH.

Alam, M. M. D. (2018). The effect of service quality on customer loyalty of generation Y â€ an empirical investigation on superstores in Bangladesh. *Malaysian Management Journal, 22*, 153–173.

Alam, S. M. S., Safiul Hoque, S. M., & Hosen, M. Z. (2010). Corporate social responsibility (CSR) of MNCs in Bangladesh: A case study on GrameenPhone Ltd. *Journal of Potuakhali University of Science and Technology*. Forthcoming, Available at SSRN: https://ssrn.com/abstract=1639570

Amor-Esteban, V., Galindo-Villardón, M. P., & García-Sánchez, I. M. (2020). Bias in composite indexes of CSR practice: An analysis of CUR matrix decomposition. *Corporate Social Responsibility and Environmental Management, 27*(4), 1914–1936.

Anandashipyard.com. (2022). *Investor relation*. Accessed May 20, 2022, from https://www.anandashipyard.com/investor-relation/

Anser, M. K., Zhang, Z., & Kanwal, L. (2018). Moderating effect of innovation on corporate social responsibility and firm performance in realm of sustainable development. *Corporate Social Responsibility and Environmental Management, 25*(5), 799–806.

Aqicn.org. (2022). *Air pollution in Dhaka, Bangladesh*. Accessed May 17, 2022, from https://aqicn.org/country/bangladesh/#:~:text=Furthermore%2C%20according%20to%20the%20most,168%20%2D%20Unhealthy%20Air%20Quality%20Index

Asiantv.com.bd. (2022). *Asian TV Live*. Accessed May 20, 2022, from https://www.asiantv.com.bd/

Bashundharagroup.com. (2022). *The Bashundhara group*. Accessed May 17, 2022, from https://www.bashundharagroup.com/article/bashundhara-group-distributes-relief-among-insolvent-families-in-bogura

Beckers, A. (2019). Towards a regulatory private law approach for CSR self-regulation? The effect of private law on corporate CSR strategies. *European Review of Private Law, 27*(2), 1–90.

Beximcopharma.com. (2022). *Corporate social responsibility-Beximco Pharmaceuticals Ltd.* Accessed May 20, 2022, from https://beximcopharma.com/about-us/corporate-social-responsibility

Beza.gov.bd. (2022). *Environmental and social management framework (ESMF) for private investment and digital entrepreneurship (PRIDE) project*. Accessed May 19, 2022, from https://www.beza.gov.bd/wp-content/uploads/2020/01/PRIDE_BEZA_ESMF_revised_Report_February-26-2020_clean.pdf

Brac.net. (2022). *Changing landscape of CSR practice in Bangladesh*. Accessed May 20, 2022, from https://innovation.brac.net/blog/changing-landscape-of-csr-practice-in-bangladesh/

Caritasbd.org. (2022). *Caritas Bangladesh*. Accessed May 20, 2022, from https://caritasbd.org/

Citigroup.com. (2022). *Audited financial statements 2020*. Accessed May 17, 2022, from https://www.citigroup.com/citi/about/countries-and-jurisdictions/data/bangladesh_audited_financial_statements_onshore_2020.pdf

Climate-transparency.org. (2022). *Bangladesh's GHG emissions per capita*. Accessed May 20, 2022, from https://www.climate-transparency.org/wp-content/uploads/2021/11/Bangladesh-CP2020.pdf

Daily-sun.com. (2022). *Bashundhara group: Rare example*. Accessed May 20, 2022, from https://www.daily-sun.com/printversion/details/481279/Bashundhara-Group:-Rare-Example-of-Corporate-Social-Responsibility

Datasoft-bd.com. (2022). *Creates customized technological solutions for changing customer needs*. Accessed May 17, 2022, from http://datasoft-bd.com/

Glonti, V., Trynchuk, V., Khovrak, I., Mokhonko, G., Shkrobot, M., & Manvelidze, L. (2020). Socialization of organization sustainable development based on the principles of corporate social responsibility. *Montenegrin Journal of Economics, 16*(1), 169–182.

Gorgenyi-Hegyes, E., Nathan, R. J., & Fekete-Farkas, M. (2021). Workplace health promotion, employee wellbeing and loyalty during COVID-19 pandemic—Large scale empirical evidence from Hungary. *Economies, 9*(2), 55.

Graafland, J. (2020). Women in management and sustainable development of SMEs: Do relational environmental management instruments matter? *Corporate Social Responsibility and Environmental Management, 27*(5), 2320–2328.

Hoque, A. S. M. M. (2018). Does government support policy moderate the relationship between entrepreneurial orientation and Bangladeshi SME performance? A SEM approach. *International Journal of Business Economics and Management Studies, 6*(3), 37–59.

Indriastuti, M., & Chariri, A. (2021). The role of green investment and corporate social responsibility investment on sustainable performance. *Cogent Business & Management, 8*(1), 1960120.

Janatabankpune.com. (2020). *Annual reports*. Accessed May 17, 2022, from https://www.janatabankpune.com/encyc/2020/10/30/2_09_30_50_Annual-Report20_1.pdf

Joseph, G., & Hashmi, A. (2018). Legitimacy and institutionalization of code of conduct: The management of business ethics. In *Research on professional responsibility and ethics in accounting*. Emerald Publishing Limited.

Kim, C., Kim, J., Marshall, R., & Afzali, H. (2018). Stakeholder influence, institutional duality, and CSR involvement of MNC subsidiaries. *Journal of Business Research, 91*, 40–47.

Kwarcinska, A. (2021). Scope of activities as a criterion determining the type of CSR actions with regard to employees: A case study. *European Research Studies Journal, 24*(4), 882–893.

Lee, J. M., & Quan, Z. (2020). A study of corporate CSR effects on corporate crisis management. *The Journal of Economics, Marketing and Management, 8*(2), 13–17.

Lu, J., Ren, L., Qiao, J., Yao, S., Strielkowski, W., & Streimikis, J. (2019). Corporate social responsibility and corruption: Implications for the sustainable energy sector. *Sustainability, 11*(15), 4128.

Manzoor, F., Wei, L., Nurunnabi, M., Subhan, Q. A., Shah, S. I. A., & Fallatah, S. (2019). The impact of transformational leadership on job performance and CSR as mediator in SMEs. *Sustainability, 11*(2), 436.

Mitu, K., Jones, N., Vintges, J., & Devonald, M. (2022). Climate risks and truncated opportunities: How do environmental challenges intersect with economic and social disadvantages for Rohingya adolescents in Bangladesh? *Sustainability, 14*(8), 4466.

Mohiuddin, M., Hossain, M., Sarker, M. N. I., Nayeem, M., Rakib, A., Islam, S., & Salehin, F. (2021). Climate change vulnerability and its impacts on live and livelihood patterns in the south-middle coastal areas of Bangladesh. In *Climate vulnerability and resilience in the global south* (pp. 487–507). Springer.

Navanapropertygroup.co.uk. (2022). *CSR policy*. Accessed May 17, 2022, from https://www.navanapropertygroup.co.uk/sustainability-and-societal-policy/

Oxfam.org. (2022). *Our goal is to create a vibrant and equitable society across Bangladesh.* Accessed May 20, 2022, from https://www.oxfam.org/en/what-we-do/countries/bangladesh

Perry, P., & Wood, S. (2019). Exploring the international fashion supply chain and corporate social responsibility: Cost, responsiveness and ethical implications. *Logistics and Retail Management, 2*(1), 77–100.

Pittaway, J. J., & Montazemi, A. R. (2020). Know-how to lead digital transformation: The case of local governments. *Government Information Quarterly, 37*(4), 101474.

Plandiv.portal.gov.bd. (2022). *Study on future direction of SMEs in Bangladesh.* Accessed May 20, 2022, from https://plandiv.portal.gov.bd/sites/default/files/files/plandiv.portal.gov.bd/notices/afbffe34_be4c_417d_b36c_ecf8db4614fc/ToR%20final%20SME.pdf

Plan-international.org. (2022). *Child marriage in Bangladesh: Findings from a national survey.* Accessed May 17, 2022, from https://plan-international.org/publications/child-marriage-in-bangladesh-findings-from-a-national-survey/#:~:text=Bangladesh%20has%20one%20of%20the,years%20and%2021%20for%20males

Radovanović, G., Miletić, L., Karović, S., Mijatović, M. D., & Bubulj, M. (2022). Influence of National Culture in companies with different ownership on employees' CSR perception in a developing country: The case of Serbia. *Sustainability, 14*(4), 2226.

Rahman, M. A., & Islam, M. A. (2019). Corporate social responsibility and sustainable development: A study of commercial banks in Bangladesh. *Applied Economics and Finance, 6*(2), 6–14.

Rahman, S., Islam, A. K. M., Saha, P., Tazkia, A. R., Krien, Y., Durand, F., Testut, L., Islam, G. M., & Bala, S. K. (2019). Projected changes of inundation of cyclonic storms in the Ganges–

Brahmaputra–Meghna delta of Bangladesh due to SLR by 2100. *Journal of Earth System Science, 128*(6), 1–11.
Rhee, Y. P., Park, C., & Petersen, B. (2021). The effect of local stakeholder pressures on responsive and strategic CSR activities. *Business & Society, 60*(3), 582–613.
Roy, T.K., & Quazi, A. (2021). *How and why do MNCs communicate their corporate social responsibility in developing countries? Evidence from Bangladesh.* Competition & Change, p. 10245294211003275.
Rasel, M., Das, D., & Khan, M. (2020). Current scenario of textile sector in Bangladesh (2019); a comprehensive review. *International Journal of Innovative Studies in Sciences and Engineering Technology, 6*(1), 52–55.
Rupalibank.org. (2020). *RupaliBankAR2020InnerWork.* Accessed May 17, 2022, from https://rupalibank.org/admin/News/RupaliBankAR2020InnerWorks.pdf
Scherer, A. G., & Voegtlin, C. (2020). Corporate governance for responsible innovation: Approaches to corporate governance and their implications for sustainable development. *Academy of Management Perspectives, 34*(2), 182–208.
Shakti.org.bd. (2022). *Shakti foundation.* Accessed May 20, 2022, from https://www.shakti.org.bd/
Siltaloppi, J., Rajala, R., & Hietala, H. (2021). Integrating CSR with business strategy: A tension management perspective. *Journal of Business Ethics, 174*(3), 507–527.
Sobhani, F. A., Haque, A., & Rahman, S. (2021). Socially responsible HRM, employee attitude, and bank reputation: The rise of CSR in Bangladesh. *Sustainability, 13*(5), 2753.
Sonalibank.com.bd. (2021). *Annual_Report_2020.* Accessed May 17, 2022, from https://www.sonalibank.com.bd/PDF_file/annualreport/2021/Annual_Report_2020.pdf
Sppl.com.bd. (2022). *Super Petrochemical Limited CSR.* Accessed May 17, 2022, from http://www.sppl.com.bd/about-us
Su, L., & Swanson, S. R. (2019). Perceived corporate social responsibility's impact on the Well-being and supportive green behaviors of hotel employees: The mediating role of the employee-corporate relationship. *Tourism Management, 72*, 437–450.
Sultana, F., (2021). *Analysis of corporate social responsibilities (CSR) and financial performance of Islami Bank Bangladesh ltd.(IBBL) in year 2015 and 2019.*
Supremepetrochem.com. (2022). *Annual report 2020–2021.* Accessed May 20, 2022, from https://www.supremepetrochem.com/pdf/SPL_AR%202020_2021.pdf
Thediplomat.com. (2021). *Who will bear the financial burden.* Accessed May 17, 2022, from https://thediplomat.com/2021/10/who-will-bear-the-financial-burden-of-supporting-the-rohingyas-in-bangladesh/#:~:text=The%20Center%20for%20Policy%20Dialogue,the%20decline%20in%20foreign%20aid
Velut, J. B., Siles-Brügge, G., & Dalingwater, L. (2021). Rethinking the dynamics of inclusion and exclusion in trade politics. *New Political Economy, 3*(8), 1–6.
Zanalaarchive.com. (2022). *Babylon scholarship program.* Accessed May 20, 2022, from http://zanalaarchive.com/babylon/csr/babylon-scholarship-program#:~:text=An%20opportunity%20for%20deprived%20but%20meritorious%20students&text=The%20scholarship%20program%20gives%20financial,financial%20condition%20of%20their%20family
Zatwarnicka-Madura, B., Siemieniako, D., Glińska, E., & Sazonenka, Y. (2019). Strategic and operational levels of CSR marketing communication for sustainable orientation of a company: A case study from Bangladesh. *Sustainability, 11*(2), 555.
Zuraida, Z., & Sugianto, S. (2021). CSR Spendings of Indonesia's IPO prospectuses. *JurnalKeuangan dan Perbankan, 25*(2), 415–433.

Afzal Sayed Munna MBA, MIM, PGCE, SFHEA, QTLS, CMBE, CMgr, MBERA, is a lecturer teaching undergraduate business courses at the University of Sunderland in London. With over 20 years' experience in media and business consultancy, both in the UK and abroad, I have been teaching in higher education institutions since 2010.

Md Sadeque Imam Shaikh, MSc, MBA, is a lecturer at the University of Wales Trinity Saint David, London and Computing and Business lecturer at Coventry University, London, UK. He has experience in publishing and consultancy in higher education.

Uzoechi Nwagbara, PhD is a professor of management at ICT University. He has BA and MA in English as well as MSc and PhD in Human Resource Management and Management, respectively, from the University of Wales, and MEd in Higher Education as well as PG Cert in Academic Practice from the University of Sunderland, UK. He is a Fellow of the Higher Education Academy (FHEA); Member Chartered Management Institute (MCMI); Member Institute of Leadership & Management (MInstLM); and Member Community Development Exchange (MCDE); and also affiliated with British Academy of Management (BAM), European Association for the Study of Literature, Culture, and the Environment (EASLCE), Postcolonial Studies Association (PSA), and Royal African Society (RAS). Dr Nwagbara has over 35 years of teaching experience and has taught in higher education for upwards of 25 years. He is external examiner for University of Cumbria and University of Bolton. He has supervised over a dozen doctoral students to completion and published well over 100 publications in high impact journals in management, leadership, CSR, sustainability, employment relations, and human resource management. Dr Nwagbara has examined PhD students internally and externally in various universities including University of Bedfordshire, Cardiff Metropolitan University, and University of the West of Scotland amongst others. He holds Visiting Teaching Positions at the University of Sunderland in London, Coventry University London, London School of Commerce, University of the West of Scotland, The University of Law UK, University of Worcester, University of Wales Trinity Saint David, and University of Plymouth as well as Visiting Professorship (management), Coal City University. He is on the editorial boards of many peer-reviewed, international journals including *Economic Insights—Trends & Challenges*, *Journal of Pan African Studies* and *Journal of Sustainable Development in Africa*.

Chapter 8
Advancing Governance Toward Sustainable Development Based on Multilateralism and Democracy: The Example of Chile

Ursula A. Vavrik

Introduction

Chile, as an OECD country, enjoys the reputation of being the most advanced country of Latin America in many respects, mostly thanks to a very liberal economy, a functioning democracy, a foreign policy based on multilateralism, and a policy system defending the rule of law and peace, among others.

Even though right wing governments followed left wing governments or vice versa on a regular basis, these basic governance rules have been respected throughout the last decades after the fall of the Pinochet regime. In comparison to many other countries in the region, Chile was able to maintain a stable economic and policy environment which finally led to the situation where Chile has become one of the most reliable partners for Western economies from that region. Under the premises that a high level of multilateralism and democracy can be considered as cornerstones of a governance system toward sustainable development, we wanted to describe and analyze these two themes in a more in-depth manner in order to find some responses to the relative success of Chile's sustainable economic and political system, not so often seen amongst other regional actors.

U. A. Vavrik (✉)
NEW WAYS Center for Sustainable Development, Vienna, Austria

U. Nwagbara et al. (eds.), *Corporate Social Responsibility Disclosure in Developing and Emerging Economies*, CSR, Sustainability, Ethics & Governance, https://doi.org/10.1007/978-3-031-61976-2_8

Cherishing Multilateralism: Chile as a Regional Leader Regarding Policies Toward Sustainable Development

One year after the new government under President Boric has been in power, multilateralism figures very high on the foreign policy agenda as one out of five priorities. Among others, Chile intends to fully participate in international organizations and commits to the achievement of the 2030 Agenda and the UN Sustainable Development Goals (MINREL, 2023a, b).

While regional integration in Latin America only advances very slowly, with respect to Mercosur or the Andean Pact in particular, fostering regional integration in Latin America represents another political priority of the new Chilean government. In that vein, as a strong symbolic step, the government signed the Escazú Agreement, as part of its green agenda, the first and only Latin American and Caribbean Environmental Agreement, only a few days after taking power (Gobierno, 2023).

Besides the 26 Free Trade Agreements Chile signed with 64 countries (EC/CD, 2023), Chile has developed more intense relationships going much further than pure trade agreements with several regions of the world. With the European Union, for example a new Advanced Framework Agreement, the most prominent agreement probably of that kind, has been elaborated and politically agreed upon recently. The relationship with Asia led to the adherence to the Pacific Alliance and in February 2023 to the Trans-Pacific Partnership (now Comprehensive and Progressive Agreement for Trans-Pacific Partnership) (EU/CD, 2023).

The close cooperation of Chile with the United Nations (UN) and the European Union (EU) is proof of the strong commitment to multilateralism and shared values with the UN and the EU such as democracy, human rights, rule of law, justice, freedom, transparency, protection of the environment, equality, equal opportunities, and peace. Chile is also very active as a candidate for membership and leading functions within the UN fora. As such, Chile just took over the presidency of ECOSOC (United Nations, 2023).

UN Cooperation Framework UN: Chile 2023–2026

In April 2023, the new cooperation framework between the United Nations and Chile was presented for the period 2023–2026 (United Nations 2, 2023). In this agreement, the government of Chile confirmed to cooperate with the UN agencies focussing on four priorities to advance the sustainability agenda in Chile.

** *Substantial equality in access and enjoyment of economic, social, cultural, and environmental rights.*

** *Strengthened democratic institutionalization and security for a new inclusive social contract.*

** *An inclusive and resilient model of sustainable development to front the climatic crisis, the loss of biodiversity and contamination.*
** *Gender equality.*

Within this cooperation, further themes of importance are migration—whereas raised immigration from especially Venezuela, Columbia, and Haiti had a negative impact on societal issues, decentralization, transitional justice, oceans, as well as healthy nutrition and food sovereignty. Furthermore, a sustainable development strategy for the Auracania region is foreseen. This region is a historically and politically sensitive area since the indigenous Mapuche population has claimed land rights over the years, a situation that has led to lasting conflicts with private owners and governments (United Nations 2, 2023).

Escazú Agreement

This Regional Agreement on Access to Information, Public Information and Justice in Environmental Matters in Latin America and the Caribbean is the first binding regional environmental agreement of this kind and the first in the world containing provisions on environmental human rights defenders, stemming from a UN conference on Sustainable Development. It entered into force in April 2021 and aims at guaranteeing the "*full and effective implementation of the rights to access environmental information, public participation in the environmental decision-making process, and access to justice in environmental matters, and the creation and strengthening of capacities of cooperation, contributing to the protection of the right of every person of present and future generations to live in a healthy environment and to sustainable development.*" Signed by 24 states, today it counts 15 state parties including Chile (Observatory, 2023; Geneva Network, 2023).

Mercosur

The Common Market of the South/Mercado Común del Sur, short Mercosur, was established in 1994 by Argentina, Brazil, Paraguay, and Uruguay initially as a trade agreement. Even though progress is rather slow, Mercosur offers a great opportunity to advance on a common sustainability agenda in the region. Today, as outlined in the Joint Communication of the Member States from July 04, 2023, on the occasion of the 63 Social Summit, the coalition of States encompasses 12 Member States including Chile as Associated Member (Mercosur, 2023), and has widened its objectives to a political agreement promoting among others the protection of human rights, democracy, and sustainable development objectives. Furthermore, advancement of gender issues, harmonization of norms for matrimony and personal relations within them, University cooperation and cultural cooperation, action against transnational crime or illicit fabrication or trade of arms figure are on the agenda (Mercosur, 2023).

Trans-Pacific Partnership Agreement

The Comprehensive and Progressive Agreement for Trans-Pacific Partnership/CPTPP, also known as TPP-11, is a free trade agreement between 11 countries around the Pacific Rim providing for almost complete liberalization of tariffs amongst participants. It comprises Canada, Mexico, Peru, Chile, New Zealand, Australia, Brunei, Singapore, Malaysia, Vietnam, and Japan. While the USA withdrew in 2016, negotiations with the UK were finalized in March 2023. In February 2023, Chile officially entered the Comprehensive and Progressive Trans-Pacific Partnership agreement. Its members also have trade agreements with the EU (see FFPortal, 2023; Institute, 2023).

Andean Pact

The original Andean Pact was founded in 1969 by Bolivia, Chile, Colombia, Ecuador, and Peru with the aim to foster regional integration and improving the quality of life of 115 million citizens. While Chile pulled out of this Pact in 1976, in 2022 President Boric considered strengthening this and other integration mechanisms such as the Andean Pact and the Community of Latin American and Caribbean States (CELAC). It is interesting to note that in 2022 the Colombian President proposed to install a common Latin American currency like the Euro in the European Union, an idea considered too premature by Boric with respect to the current status of regional integration. (Comunidad, 2023; Bloomberg, 2023).

Host of the Secretariat of the UN BBNJ Treaty: Ocean Governance

Chile's engagement in multilateralism and ocean governance is further demonstrated by the proposal made in March 2023 to host the UN Secretariat of the Biodiversity Convention of Natural Areas Beyond Jurisdiction (BBNU). In this context, work should be done "*on the regulation and protection of the marine biodiversity in areas and spaces that are outside national jurisdiction*." The Treaty text has been adopted in June 2023 and will be open for signature from September 2023 to September 2025 (MINREL 2, 2023; United Nations 3, 2023).

Antarctic Treaty

This multilateral instrument for the preservation of peace and cooperation on the continent, entered into force 62 years ago and counts 29 consultative countries in 2023. In 2021, the government of Chile proposed the adoption of a new Antarctic Law that entered into force in the same year, and the National Antarctic Tourism Policy. The impact of climate change and the importance of Antarctica for the analysis of this phenomenon were prioritized (MINREL 3, 2023).

Advanced Framework Agreement EU: Chile

The European Union signed its first bilateral agreement, an association agreement, with Chile in the year 2002 (EU-Chile Agreement), having elected Chile as a strategic partner. 20 years later, a political agreement for a modernized new Advanced Framework Agreement (AFA) between the European Union and Chile has been confirmed between the two parties during a visit of Mrs. Von der Leyen, President of the European Commission, in mid-June 2023. The AFA "*provides for a modernized comprehensive legal framework for EU-Chile relations and replaces the current Association Agreement.*" In addition to a trade chapter, it includes numerous articles on extensive political cooperation which is particularly relevant also for the sustainability agenda. It is the first international agreement of that kind with a huge number of environmental and sustainability clauses. Once implemented it can be considered as one of the most advanced bilateral agreements. It "*paves the way for enhancing political, regional and global cooperation between two like-minded partners. "Further, it covers the EU standard clauses on human rights, International Criminal Court (ICC), Weapons of Mass Destruction (WMD), Small Arms and Light Weapons (SALW), and counter-terrorism. It also encompasses cooperation in areas such as health, environment, climate-change, ocean governance, energy, tax, education and culture, labour, employment and social affairs, science and technology, and transport. The AFA further addresses legal cooperation, rule of law, money laundering and terrorist financing, organized crime and corruption.*" It especially also puts a greater emphasis on sustainable development issues (EU/CD, 2023).

Most of the above-mentioned agreements have a significant influence on national policy, the private sector, and particularly on sustainable development policy. They shape the political dialogue, and the attention given both in the media, at university level or in society as a whole. They steer the country toward internationally agreed higher standards, stimulating both national, private sector and civic initiatives. Chile, e.g., has developed a comprehensive renewable energy policy including the installation of a huge solar power capacity, and has gained reputation as a leader in green hydrogen. With about 20% of its territory dedicated to national parks or protected areas (Ecocamp Patagonia, 2023), and a very advanced ocean governance strategy, Chile stands out as a regional leader, together with Costa Rica. The adherence to several important international conventions and agreements, especially the multi-faceted new Advanced Framework Agreement with the European Union will provoke a further step toward governance for sustainable development and a more sustainable future (see also European Commission, 2023).

Advancement Toward Sustainable Developemen at Nationale Level - Legal Framework and Precursor Activities Chile: Leader in Sustainability

With 78,2 points Chile has again become the regional leader in the SDG dashboard ranking in 2023 (SDG Rankings, 2023). In the same vein, the most recent Longitudinal Study 2023, a comparative perception study conducted by Fundación Imagen de Chile in 12 priority markets analyzed, describes Chile as the regional leader in comparison to Argentina, Brazil, Peru, and Colombia within two out of three classifications (Imagen, 2023):

** *Development of sustainability and environmental protection (43%)*
** *Innovation and entrepreneurship (40%, second rank)*
** *Scientific and astronomical development (46%)*

Whereas some years ago Chile was still considered lagging behind in many environmental and sustainable development policy areas, Chile has advanced nowadays to a regional leader with respect to the installed legal framework and public and private sector precursor initiatives at national level (see also COES, 2018). In particular, the build-up of renewable energy installations such as solar and wind power, as well as a green hydrogen strategy and advancements in climate policy are of relevance. Chile aspires to be carbon-neutral in 2050, and disposes of a remarkable renewable energy strategy, which could serve as a model for many Latin-American countries. Only a few other countries such as Costa Rica and Paraguay have better records in this area (Weforum, 2023). The "Cerro Dominador Solar Power Plant," a solar thermal tower with 110 MW, was inaugurated in June 2021 and at the time the largest of its kind with storage in Latin America, and the third largest energy storage plant worldwide (Wikipedia, 2023). With its 10,800 mirrors of 140 m^2 each, it has the capacity to avoid about 640,000 tons of CO_2 emissions per year (Laderasur, 2021).

With respect to astronomical development, the biggest telescopes for the observation of the cosmos are placed in Chile in the Atacama Desert. For the time being the ESO Parranal Telescope considered to be the biggest Very Large Telescope (VLT) worldwide collects data with four unit telescopes of 8,2 m diameter each (ESO, 2023), attracting researchers from all over the world. One only realizes the light contamination the World has been suffering so far after having experienced a visit to these huge observatories in Chile where the Milky Way can be seen with mere eyes, without any telescope almost every night. One can even take photos with the iPhone of the Milky Way, which is a seldom phenomenon and a possibility Chile is offering as long as light pollution will not advance too far. The local astronomical museum, therefore, warns of a very obvious risk that without stopping worldwide light pollution, Chile and the world would lose this window to the cosmos which would be most deplorable. Experts therefore call for an international convention to protect relevant areas and reduce light pollution to more sustainable levels, which is considered being close to a point of no return. In this respect, since light can harm humans, wildlife, and our climate, more than 200 dark sky parks have been established all over the world so far (Dark Sky, 2023), preserving the dark sky. While in 2001 Flagstaff, Arizona, was called the first international dark sky city (Dark Sky 2, 2023), in 2015, the "Gabriela Mistral Dark Sky Sanctuary" in Chile was announced as the first international dark sky sanctuary. The site of the Association of Universities for Research in Astronomy (AURA) in the Elqui Valley of northern Chile got its name in honor of Nobel laureate Gabriela Mistral stemming from that region (Dark Sky 3 2023).

National Strategy for the Implementation of the Agenda 2030 (Gobierno 2, 2023)

In October 2022, Chile started a participatory and inclusive process aimed at the elaboration of the National Strategy for the Implementation of Agenda 2030, which was presented in early 2023.

This Strategy is based on five strains:

- ***1. People:*** *Focussing on the challenges related to the reduction of certain social vulnerabilities and on strengthening the social system.*
- ***2. Planet:*** *Addressing the environmental crisis, highlighting the necessity to transform the relationship between humans and nature in order to reach a transition to a just socioecology. Themes like water, sanitation, sustainable cities sustainable consumption and production, climate action, submarine life, life on earth and peace, justice, and institutionally are analyzed.*
- ***3. Prosperity:*** *Touching upon challenges like productivity, innovation, infrastructure, territorial equity, gender equality, and decent work.*
- ***4. Peace, justice, institutions:*** *Stressing the importance of promoting peaceful, just and inclusive societies, treating themes like civil security, violence, and discrimination.*
- ***5. Alliances:*** *Distinguish the importance to establish collaborative relations in order to tackle problems and challenges together. Themes like financing, access to technologies, institutional cohesion, alliances of all kinds, also those to achieve the SDGs.*

Consequently, the objectives for strain 1 are oriented toward an improved social, health and educational system, the reduction of inequalities, as well as toward the amelioration of living and housing conditions.

In strain 2, Chile aspires to advance toward a just water transition, to promote the prevention and decontamination, especially regarding air quality, to promote sustainable production and consumption patterns as well as decoupling economic growth from environmental degradation. Furthermore, the circular economy will be stimulated, and institutionalization and public policies for biodiversity enhanced. Thanks to the Escazú Agreement, access to information, public participation, and justice in environmental matters will be promoted.

Moreover, in strain 3, policies to reduce inequality in all its expressions are pursued, as well as policies to ensure quality employment and sustainable housing and sustainable cities. The digital divide needs to be diminished and models of sustainable and inclusive production are to be envisaged.

In strain 4, civil security shall be enforced through prevention of violence and the promotion of a culture of peace. All kinds of discrimination are to be eradicated by promoting inclusive policies and sensitization for human rights. Decentralization shall be strengthened as well as civic participation in decision-making. Corruption will need to be combatted, and international cooperation and solidarity against violence, discrimination, and poverty are to be enhanced.

Finally, in strain 5, objectives relate to the identification of the financial resources designated to public action oriented toward the achievement of the SDS, of the existing gap, and the incorporation of the SDGs in management and budget cycles. The concept of "Development in Transition" is to be promoted at the international level. The national adaptation capacity to technological transformation shall be raised, and export with technologies entailing a better-added value for the offer of the country shall be supported. International cooperation regarding the advancement toward achievement of the SDGs shall be reinforced (Gobierno 2, 2023, shortened and translated from the Spanish version of the strategy).

It would be beyond the scope of this chapter to analyze the achievements of the SDGs with current policy measures. This remains to be done at a later stage. While some achievements definitely point in the right direction such as the climate law, and the new constitution, in other areas such as water management, desertification, efficient education, health and social policy, the implementation of relevant SDG goals still remains to be improved.

Legal Framework for Environmental and Sustainability Policy and Some Relevant Initiatives

Chile disposes of several new legislative and strategic documents such as the new climate law (Gobierno, 2022), the new strategy for renewable energy policy (Gobierno, 2021), and a new Green hydrogen strategy. Moreover, a Natural Capital Committee has been put in place (Gobierno 3, 2023).

According to experts, Chile could be a world leader in green hydrogen production and export, thanks to its geographical situation. While in the north of Chile solar energy can be produced in huge quantities in the Atacama Desert, wind energy prevails in the south of Chile, with a precursor plant already in place at Punta Arenas, making Chile one of the best-positioned countries worldwide for hydrogen investment and development. Unfortunately, not everything is developing as smoothly as one would expect. It appears that several solar and production companies are not able to feed their electricity into the national net, due to the lack of sufficient transmission capacity of the national electricity net, thus entailing huge excess quantities and losses of electricity (see also Axlegal, 2022). The government has been working on a solution for this unsustainable situation mainly by improving storage capacity (see Axlegal, 2022).

There exist both projects for a green hydrogen powered train (Invest Chile, 2022) and green hydrogen powered buses. At Rancagua, about 80 km south of Santiago, the first Chilean and also Latin American electric bus factory, Reborn Electric Motors has emerged in July 2022, disposing of a 100% electric bus fleet. Around 200 zero-emission buses can be produced yearly, thus contributing to the decontamination of cities and the reduction of 65,000 t CO_2 emissions to the atmosphere per year (Chile, 2022). The first electric buses were introduced in Santiago in 2017, having reached the landmark figure of 1000 units in February 2023. The government's goal now is to double the electric bus fleet by the end of the year and to reach a completely electric bus fleet by 2040. So far, the electric bus fleet of Chile is the biggest worldwide outside China (Bnamericas, 2022).

In January 2023, President Boric launched the Natural Capital Committee to promote the sustainable development of Chile. "*The committee's mission will be to advise and propose actions to the president in matters related to measuring, valuing, protecting, restoring and improving Chile's natural capital. Amongst other things, the committee will design the future implementation of a national environment accounts system based on ecosystem assets and aligned with criteria published by the UN*" (Gobierno 3, 2023).

While, in several respects, Chile belongs to the group of most advanced economies, we may just touch upon some themes with less positive impacts such as the underdeveloped social system - another task the new government is working on- or the problems with desertification, water access and availability, land and forest use as well as land distribution and exploitation.

From an institutional governance perspective, one notes a high rate of personal changes not only in political positions but also in public service positions when governments are changing, last but not least in schools and hospitals. This situation is unusual in European or other Western economies and entails very often inefficiency, lack of knowledge transfer, diminished service quality, lost opportunities and unsustainable conditions, expressed in a more general way. For a more sustainable public sector policy these issues will need to be taken into account in the near future.

The New Constitution: Governance Towards Strengthening Democracy, Enhanced Sustainability, and Social Coherence

Chile as an OECD country is the most advanced liberal economy amongst Latin American States. Very much aligned to the liberal system of the United States of America, the social security net has not been that far developed as, e.g., in European States which has entailed inequalities in several respects such as access to health, education, or transport systems. In 2019, therefore, a slight rise in the price of a metro ticket led to a series of demonstrations and riots, a situation that made history as the "estallido social" (social explosion). Whereas the demonstrating population initially claimed a more just social system, a multiparty parliamentary decision finally calmed down the situation with the announcement of a new constitutional process. The decision and elaboration of this new constitution can be described as a path breaking governance process, based on the promise to renew the constitution of the Pinochet era with a more democratic and modern one. It was to be elaborated from scratch ("hoja blanca") in a democratic, inclusive, and participatory process, with parity for men and women in the constitutional committee. Already in this first phase, even though it was rejected later, it focussed on establishing a „green and sustainable "constitution, strengthening" *the potential role of direct and participatory democratic institutions, and how to improve democratic representation, especially of women, indigenous communities, and other historically under-represented minorities"* (David Rockefeller Center, 2022).

The social unrest also led to a shift to a leftwing government in 2021. During the presidential elections of March 2021, the two moderate candidates representing the center of the population lost the pre-selections and two relatively "extreme" candidates, from both the far right wing and the left wing parties succeeded in entering the final round, in which President Gabriel Boric, a former student leader representing the far left, finally won. This new leftist government under President Gabriel Boric has since then promised many reforms as already suggested longer ago by the OECD, such as enhancing social cohesion, a more sustainable economic development, respect of human rights, diversity, and minorities, as well as women's rights. The overall aim was to build a fairer society and to shift slowly the institutional system more toward a European-like welfare state, in which health, education, and a new pension system are regarded as basic rights and are not defined as goods that depend on financial wealth.

In 2020, in the constitutional plebiscite, 80% voted in favor of electing a Constitutional Convention (Espacio Público, 2023). Thus, in an obligatory election process Chileans chose their representatives for the 155 persons committee which should elaborate the new constitutional charter. Whereas this format allowed all kinds of candidates with most diverse professional backgrounds to assure a good representation of the multifaceted society, including the indigenous population, it was also the first time that nontraditional candidates were represented in national politics, directly compensating for the lack of historical representation for various groups. The constitutional text "*included five significative, innovative topics: A welfare state, decentralization of power (both political and economic), from national to regional and local institutions, environmental protection, rights and recognition for indigenous peoples, and parity between women and men"* (Espacio Público, 2023).

However, the process failed in the first place due to a multitude of internal and external factors (see Espacio Publico, 2023). Among others, delegates did not show enough flexibility for compromises and overloaded the constitution text. Instead of a slim, persuading, and practical new constitution, the final text included too many controversial, not practical and not easily implementable clauses. In the referendum of September 2022 therefore, 63% of the population rejected this new constitutional text (America's Quarterly, 2023). Whereas during the whole process a "No" was considered almost catastrophic for the further political and economic development of Chile, reality has shown that this was not the case. On the contrary, from a democratic governance perspective, the constitutional process has revealed that Chile was quite successful in handling a difficult political situation without too many negative side effects in a rather calm and controlled way.

After the rejection of the first new constitutional text, a second process was installed by the Parliament, similar to the first one, but revised: First, a new text was to be prepared by a team of experts and then transmitted to elected delegates for fine-tuning. Second, the number of delegates was reduced to 50, and they were bound to a set of 10 principles. Thus, in a second step, in January 2023 a group of experts nominated by the Congress started to draft a new constitutional text, and in June 2023, a committee of 50 delegates was elected again in a mandatory election. This time the right wing parties reached a majority. In December 2023, this text will then be again subject to adoption by the Chilean population through a referendum. By the time of writing this chapter, in July 2023, according to polls, more than 80% of the population backed the text as proposed by the team of experts. The text as finalized by the elected delegates included, amongst others, a new clause proposes to limit the entrance into parliament to parties with a minimum of 5% of votes as it is the case in most Western parliamentary systems. So far, the limit was much less, about 2%, which facilitated a huge number of political parties getting seats in the Parliament, which was rather counterproductive from a governance perspective. With actually 21 political parties it has been rather difficult to find democratic compromises (see also Klein, 2022).

Conclusion

Having described and analyzed, especially the two cornerstones of sustainable development, multilateralism, and democracy, one may conclude that Chilean governments safely governed their country through the last years while standing out as a regional leader in terms of both strong defenders of multilateralism and democracy. The new government of President Boric even tries to accelerate this process by an even stronger dedication to multilateral processes, also with respect to regional integration. The example to guide the population toward a new constitution has been particularly interesting from both a governance perspective and a sustainability perspective. Even if the second proposal was rejected again in the following referendum, the democratic process leading to this new constitution was highly remarkable from a governance perspective. Thus, thanks to a policy direction striving for strengthened multilateralism and democracy, Chile has advanced to a regional leader in many respects, last but not least in governance for sustainability.

References

America's Quarterly, Reaction: Chile's new constitutional roadmap. (2023, January 11). *Amercia's quarterly.* Accessed July 10, 2023, from https://www.americasquarterly.org/article/reaction-chiles-new-constitutional-roadmap/

Axlegal, Chile's Renewable Boom Has A Transmission Problem. (2022, November 29). *Axlegal.* Accessed Sep 17, 2023, from https://ax.legal/2022/11/29/chiles-renewable-boom-has-a-transmission-problem/

Bloomberg, 2023, Chile's Boric Says Too Soon to Adopt Colombia's Proposal for LatAm Common Currency. (2023, August 9). Bloomberg. Accessed July 09, 2023, from https://www.bloomberglinea.com/english/chiles-boric-says-too-soon-to-adopt-colombias-proposal-for-latam-common-currency/

Bnamericas, Santiago looking to double electric bus fleet after hitting 1000 units milestone. (2022, February 2). *Bnamericas.* Accessed July 17, 2023, from https://www.bnamericas.com/en/news/santiago-looking-to-double-electric-bus-fleet-after-hitting-1000-unit-milestone

Chile, Creating Future, Discover the first electric buses made in Chile. (2022, September 30). *Chile.* Accessed July 26, 2023, from https://marcachile.cl/en/innovation-entrepreneurship/discover-the-first-electric-buses-made-in-chile/

COES, From green laggard to regional leader: explaining the recent development of environmental policy in Chile. (2018). *COES.* Accessed July 08, 2023, from https://coes.cl/publicaciones/from-green-laggard-to-regional-leader-explaining-the-recent-development-of-environmental-policy-in-chile/

Comunidad Andina, Quiénes somos. (2023). *Comunidad.* Accessed July 08, 2023, from https://www.comunidadandina.org/quienes-somos/

Dark Sky, All dark sky places. (2023). *Dark Sky.* Accessed July 09, 2023, from https://darksky.org/places/

Dark Sky, International dark sky places. (2023). *Dark Sky 2.* Accessed July 09, 2023, from https://darksky.org/what-we-do/international-dark-sky-places/

Dark Sky, Chilean Astronomicla Site Becomes World's First International Dark Sky Sanctuary. *Dark Sky 3.* Accessed July 09, 2023., from https://darksky.org/news/chilean-astronomical-site-becomes-worlds-first-international-dark-sky-sanctuary/

Davon Rockefeller Center for Latin American Studies, Harvard University, academic forum for the new constitution in Chile. (2022). *David Rockefeller Center.* Accessed July 18, 2023, from https://drclas.harvard.edu/chile-en

Ecocamp Patagonia, The Second Largest National Parc in Chile, (2023). *Ecocamp.* Accessed Sep 17, 2023, from https://www.ecocamp.travel/blog/kawesqar-national-park#:~:text=4%2C5%20million%20hectares%20are,the%20country's%20territory%20is%20protected

Espacio Público, A new constitution for Chile: Let's try again. (2023, February 21). *Espacio Público.* Accessed July 10, 2023, from https://espaciopublico.cl/a-new-constitution-for-chile-lets-try-again/

European Commission, Proposal for a Council Decision on the conclusion, on behalf of the European Union, of The Advanced Framework Agreement between the European Union and its Member States, of the one part, and the Republic of Chile, of the other part, Brussels, COM/2023) 432 final. (2023, July 05). *EC/CD*. Accessed July 14, 2023, from https://eur-lex.europa.eu/resource.html?uri=cellar:bcfd4c45-1b0f-11ee-806b-01aa75ed71a1.0001.02/DOC_1&format=PDF

European Commission, The EU-Chile Agreement explained. (2023). *European commission*. Accessed July 05, 2023, from https://policy.trade.ec.europa.eu/eu-trade-relationships-country-and-region/countries-and-regions/chile/eu-chile-agreement/agreement-explained_en

European Space Observatory, Parranal Telescopes. (2023). *ESO*. Accessed Sep 17, 2023, from https://www.eso.org/sci/facilities/paranal/telescopes.html

Freshfruitportal, Chile formally joins Transpacific Partnership. (2023, March 02). *FFPortal*. Accessed July 05, 2023, from https://www.freshfruitportal.com/news/2023/03/02/chile-formally-joins-trans-pacific-partnership/

Geneva Environment Network, A Step forward in Environmental rights in Latin America and the Caribbean; 1st Meeting of the Conference of the Parties to the Escazú Agreement in Santiago de Chile, 20–22 April 2022. (2023, April 19). *Geneva network*. Accessed July 17, 2023, from https://www.genevaenvironmentnetwork.org/events/a-step-forward-for-environmental-rights-in-latin-america-and-the-caribbean-1st-meeting-of-the-conference-of-the-parties-to-the-escazu-agreement-in-santiago-de-chile-20-22-april-2022/

Gobierno de Chile, Minister Jobet announces 66 measures to guide the development of Chile's energy sector over the next decades. (2021, October 21). *Gobierno*. Accessed Sep 17, 2023, from https://www.gob.cl/en/news/minister-jobet-announces-66-measures-to-guide-the-development-of-chiles-energy-sector-over-the-next-decades/#:~:text=100%25%20zero%2Demission%20energy%20used%20in%20electricity%20generation%20by%202050.&text=Renewable%20energies%20to%20contribute%2080%25%20of%20electricity%20generation%20by%202030.&text=100%25%20access%20to%20electricity%20for,emphasis%20on%20the%20indigenous%20population

Gobierno de Chile, A milestone in Chile's environmental history from today we have our first framework law on climate change. (2022, June 13). *Gobierno*. Accessed Sep 17, 2023, from https://www.gob.cl/en/news/a-milestone-in-chiles-environmental-history-from-today-we-have-our-first-framework-law-on-climate-change/

Gobierno de Chile, Why is it so important for Chile to be part of the Escazú Agreement. (2023, April 20). Gobierno. Accessed July 02, 2023, from https://www.gob.cl/en/news/why-it-so-important-chile-be-part-escazu-agreement/

Gobierno de Chile, Estrategia nacional para la implementación de la agenda 2030, shortened and translated into English. (2023). *Gobierno 2*. Accessed July 14, 2023, from https://www.chileagenda2030.gob.cl/storage/docs/Estrategia_de_Implementacion_Agenda2030.pdf

Gobierno de Chile, President Boric Launches Natural Capital Committee to Promote Sustainable Development in Chile. (2023, January 17). *Gobierno 3*. Accessed July 18, 2023, from https://www.gob.cl/en/news/president-gabriel-boric-launches-natural-capital-committee-promote-sustainable-development-chile/

Imagen de Chile, Chile leads in Sustainability, environment and scientific development in the region. (2023, Julio 14). *Imagen*. Accessed July 20, 2023, from https://marcachile.cl/en/imagen-de-chile/chile-leads-in-sustainability-the-environment-and-scientific-development-in-the-region/

Institute for Government, The Comprehensive and Progressive Agreement for Transatlantic Partnership. (2023, April 13). *Institute*. Accessed July 07, 2023, from https://www.instituteforgovernment.org.uk/explainer/comprehensive-and-progressive-agreement-trans-pacific-partnership#:~:text=The%20CPTPP%20is%20a%20free,%2C%20Malaysia%2C%20Vietnam%20and%20Japan

Invest Chile, What Will The First Chilean Green Hydrogen-Powered Train Be Like?. (2022, January 13). *Invest chile*. Accessed July 20, 2023, from https://blog.investchile.gob.cl/green-hydrogen-train-created-in-chile

Klein Fabian et al. (2022). Chile report, sustainable governance indicators 2022. *Klein*. Accessed July 17, 2023, from https://www.sgi-network.org/docs/2022/country/SGI2022_Chile.pdf

Laderasur, Recorriendo Cerro Dominador: La única planta termosolar que funciona en medio del desierto de Atacama. (2021, Septiembre 7). *Laderasur*. Accessed July 10, 2023, from https://laderasur.com/articulo/recorriendo-cerro-dominador-la-unica-planta-termosolar-que-funciona-en-medio-del-desierto-de-atacama/

Mercosur. (2023). *Mercosur*. Accessed July 04, 2023, from https://www.mercosur.int

MINREL, Ministerio de Relaciones Exteriores, Minister Alberto van Klaveren Highlights Ministry's Achievements at the Participative Public Account 2022–2023. (2023a, 17 May). *MINREL*. Accessed July 02, 2023, from https://www.minrel.gob.cl/news/minister-alberto-van-klaveren-highlights-ministry-s-achievements-at-the

MINREL, Chile formalizes its proposal to host the secretariat of the United Nations BBNJ treaty. (2023b, March 9). *MINREL 2*. Accessed July 07, 2023, from https://www.minrel.gob.cl/news/chile-formalizes-its-proposal-to-host-the-secretariat-of-the-united

MINREL, Chile participates in the Antarctic Treaty Consultative Meeting. (2023, June 14). *MINREL 3*. Accessed Jul 23, 2019, from https://www.minrel.gob.cl/news/chile-participates-in-the-antarctic-treaty-consultative-meeting

Observatory on Principle 10. (2023). *Observatory*. Accessed Jul 25, 2023, from https://observatoriop10.cepal.org/en/treaty/regional-agreement-access-information-public-participation-and-justice-environmental-matters

Sustainable Development Report, Rankings. (2023). *SDG Rankings*. Accessed Jul 08, 2023, from https://dashboards.sdgindex.org/rankings

United Nations, New Economic and Social Council leadership takes the helm. (2023, July 27). *United Nations*. Accessed Sep 17, 2023, from https://news.un.org/en/story/2023/07/1139202

United Nations, Marco de cooperación de las naciones unidas para el desarrollo sostenible en Chile 2023–2026. (2023, April 2). *United Nations*. Accessed July 01, 2023, from https://chile.un.org/es/227196-marco-de-cooperaci%C3%B3n-de-las-naciones-unidas-para-el-desarrollo-sostenible-en-chile-2023-2026

United Nations, Intergovernmental Conference on Marine Biodiversity of Areas Beyond National Jurisdiction. (2023, September 3). *United Nations*. Accessed Sep 17, 2023, from https://www.un.org/bbnj/content/sign-agreement

Wikipedia, Cerro Dominador Solar Termal Plant. (2023). *Wikipedia*. Accessed July 18, 2023, from https://en.wikipedia.org/wiki/Cerro_Dominador_Solar_Thermal_Plant

World Economic Forum, Latin America, How Chile is Becoming a Leader in Renewable Energy. (2023, January). *Weforum*. Accessed Jul 09, 2023, from https://www.weforum.org/agenda/2023/01/how-chile-is-becoming-a-leader-in-renewable-energy/

Ursula A. Vavrik, PhD, holds a Master's degree and a PhD (1990) in Economic and Social Sciences and has worked on environment and sustainable development issues for over fifteen years. She has a wide-ranging career history encompassing EU institutions, multinational organizations (the UN and OECD), academia, the private sector, and NGOs. Having begun as a journalist in1985, she then worked as a consultant for UNDP, UNEP, and NGOs and was project coordinator in Kenya for four years. Her EU expertise comes from postings as environment attaché at the Austrian Permanent Representation (1997-1999), spanning Austria's 1998 EU Presidency, as national expert at DG Environment for three years, and as lecturer in European Politics at the Paris Institute of Political Studies ('Sciences Po'). In the year 2000, she was appointed guest professor for environmental management at Vienna's University of Agriculture and has held classes on sustainable development at the Vienna University of Economics for over ten years. During the Austrian EU Presidency in 2006, she served as a senior advisor on development at the Austrian Ministry of Foreign Affairs. Most recently, she has become EU Policy Director at the EEB. She is married and has two children.

Chapter 9
Corporate Social Responsibility (CSR) and Sustainable Development: Exploring the Nexus in the Sierra Leonean Context

Henry Yamba Kamara, Uzoechi Nwagbara, Chigozie A. Ugwoji, and Ngozi Ibeawuchi

Introduction

Corporate Social Responsibility (CSR) is considered as obligation of organisations to the environment and society beyond what the law stipulates (Crane & Matten, 2010; Visser, 2006; Carroll, 1979, 1991). It also refers to an organisation's commitment to practise social and environmental sustainability so as to be a good corporate citizen in the environment, which it operates (Banerjee, 2008). CSR is often used by firms to court legitimacy (Deegan, 2002), increase the level of trust (Belal, 2008), decrease stakeholders' criticisms (Crowther & Aras, 2008), manage reputation (Ferguson et al., 2016) and increase "social licence" to operate (Ferguson et al., 2016). For decades, companies have engaged in CSR, yet one question that persists is the impact of CSR initiatives on the people and their environment (Idemudia, 2010; Ite, 2004) as well as the impact of their operation on sustainable development

A major portion of this chapter is reused from Dr. Henry Yamba Kamra's PhD thesis, which was awarded by Cardiff Metropolitan University in October 2020. Permission to reuse some of the contents of the PhD thesis has been sought and granted.

H. Y. Kamara
Smallholder Commercialisation and Agrobusiness Development Project (SCADeP), Freetown, Sierra Leone

U. Nwagbara · C. A. Ugwoji (✉)
University of Sunderland in London, London, UK
e-mail: uzoechi.nwagbara@sunderland.ac.uk; Chi.ugwoji@sunderland.ac.uk

N. Ibeawuchi
University of Greenwich, London, UK
e-mail: Ngozi.ibeawuchi@coventry.ac.uk

U. Nwagbara et al. (eds.), *Corporate Social Responsibility Disclosure in Developing and Emerging Economies*, CSR, Sustainability, Ethics & Governance, https://doi.org/10.1007/978-3-031-61976-2_9

(Crowther et al., 2019). Specifically, CSR remains a problematic issue in developing countries (Agbiboa, 2012).

Despite attempts made globally to encourage corporations to engage in CSR (KPMG, 2017), research shows that the processes through which organisations engage in CSR are undermined by managerial ideology and orientation, which elicit stakeholders' criticism (Frynas, 2005, 2009). Therefore, stakeholders (NGOs, the governments, the media and communities) argue that managerial orientation is largely responsible for apparent lack of social responsibility (Okoye, 2009), accountability (Siltaoja, 2009), transparency (Agbiboa, 2012), legitimacy (Deegan, 2002) and neglect of stakeholder engagement (Belal, 2008). Additionally, although approaches to CSR vary, some organisations engage in CSR as a means to advance shareholders' interest, increase profit and strategically position their business; whilst some engage in CSR initiatives out of a sense of moral obligation to the environment and society. Given that CSR is a voluntary activity, organisations tend not to be transparent, accountable and responsible in their CSR activities (Crane & Matten, 2010). This situation is more complex in developing countries, such as Sierra Leone (Porto, 2010), where there are weak institutions (Kolk & Lenfant, 2013; Maconachie, 2008, 2016), "resource curse" (Obi, 2010; Karl, 1997) issue, poor governance regime (Frynas, 2005) and leadership problems (Angus-Leppan et al., 2010), which negatively impact sustainable development. Thus, this chapter argues that the context of Sierra Leone, which is a nation reeling in the bangs and pains of unsustainable organisational practice and poor CSR, can widen perspectives as well as facilitate a better understanding of how CSR can be instrumental in sustainable development, which is an understanding crucially needed in developing countries in the wake of SDGs and their ideals (Kamara & Nwagbara, 2015).

Background and Rationale

Despite huge natural and human endowments, Sierra Leone is plagued by CSR dilemma (Porto, 2010); this situation is not unrelated to the presence of MNCs in the country, which furthers underdevelopment and socio-environmental problems (Maconachie, 2016; Christian Aid, 2013). The presence of MNCs in the global economy equation has triggered and sustained the proliferation of social, economic and environmental problems in developing countries such as Sierra Leone (Datzberger, 2014). Guided by this proposition, Deegan (2002) argues persuasively that it is crucially important to understand capitalism and MNCs' presence in the world—particularly in developing countries—for deeper knowledge of the premise of underdevelopment. This reason partly explains stakeholder pressures for businesses to rethink their roles in society (Frynas, 2005). Accordingly, Raman and Lipschutz (2010) contend that although the process of globalisation has also triggered opportunities for the MNCs from the global south as Foreign Direct Investment (FDI) has increased from a mere 5.2% in 1990 to 14.2% in 2006; it, however, remains "to be seen whether these … MNCs with their poor record of CSR would modify their behaviour" (pp. 5–6).

Studies have indicated that CSR should be conceived from the developing countries' perspective as an instrument for sustainable development (2014, Richey & Ponte, 2014; Okoye, 2009) and a win-win approach for business and society (Williams & Preston, 2018). Although an "essentially contested" concept (Okoye, 2009), scholars (Emeseh & Songi, 2014; Crowther & Aras, 2008) maintain that CSR can contribute to sustainable development as well as reduce and/or eradicate poverty, particularly in developing countries (KPMG, 2017; United Nations, 2006; United Nations Sustainable Development, 1992). For example, Zwetsloot (2003) argues that CSR appears to be a leading principle for business innovation aimed at reinventing organisational practices that can impact positively on profit, environment and the people. Accordingly, Desjardin (2000) posits that CSR is a fundamental issue in business responsibility and sustainability. International organisations such as the World Bank, United Nations and developmental agencies including the Department for International Development (DFID), consider CSR as an idea that the private sector including businesses can utilise in tackling poverty and underdevelopment and to contribute to sustainable development (Visser, 2006, 2013).

Sustainability explains the ability to maintain various processes and systems in a social, environmental and economic manner over time (Crowther et al., 2019; Elkington, 1997). Although sustainability originated in natural resource economics, it has gained traction as well as wider currency in relation to social equality and sustainable development (Bebbington & Unerman, 2008). Some organisations and economists do not subscribe to the notion of CSR, because it implies an obligation to future generations and society beyond those contained in the obligatory legal requirements of business (Crowther & Aras, 2008). Nonetheless, most businesses including Multinational Corporations (MNCs) now embrace the concept of CSR and engage in CSR initiatives. These initiatives focus on CSR and sustainability largely in relation to moral obligation, and offer insight into ethical notions relevant to environmental and economic sustainability, social equity and sustainable development.

Sustainable development (sustainability) began with the Brundtland Report that was published in 1987, and named *Our Common Future* (World Commission on Environment and Development, WCED, 1987). At its core are interdependence and multilateralism of nations in the search for a sustainable development path. Consequently, the debate regarding Africa's future, rise and sustainable development have been taken seriously recently following the publication of *Our Common Interest* (Commission for Africa, 2005), the report of the Commission for Africa. Also, in 2010 a sequel of this report, *Still Our Common Interest*, was launched to precipitate efforts by African leaders to expedite actions concertedly to convert unparalleled economic opportunities in Africa into healing social wounds, protecting the environment, advancing national development and for the foremost part advancing sustainable development on the continent.

In addition, the challenges for CSR in developing countries were further shaped by the creation of Millennium Development Goals (MDGs) in 2000 and subsequently Sustainable Development Goals (SDGs), which resonate with "a world with less poverty, hunger and disease, greater survival prospects for mothers and infants, better educated children, equal opportunities for women, and a healthier environment" (United Nations, 2006, p. 3). MDGs have been replaced by SDGs, the focus of the 2030 Agenda for Sustainable Development, which were adopted by the United Nations Sustainable Development Summit in 2015. These goals extend the work started by the MDGs that propelled a global campaign from 2000 to 2015 aimed at eradicating poverty in its various dimensions. However, whilst the MDGs essentially apply to developing countries, the SDGs apply to all UN member states universally; they are also considered to be more ambitious and comprehensive in orientation than the MDGs (KPMG, 2017). Given their power and influence, it is expected that MNCs can play a big role in actualising the ideals of SDGs, particularly in developing countries like Sierra Leone (Maconachie, 2008).

Study Context

With a population of about 6 million people, Sierra Leone has been described as a small country in West Africa where most of its people live below the poverty line (Maconachie, 2016; Datzberger, 2014). The country mainly depends on subsistence agriculture and gold mining, whilst the manufacturing sector mainly focuses on processing raw materials and light manufacturing for domestic consumption (Akiwumi, 2014). Sierra Leone's natural resources include gold, diamond, bauxite, chromite, titanium ore and iron ore. Diamond mining generates about half of the country's total exports. Also, the country's economy is hugely reliant on international aid for development and for provision of social goods. In 2006, 343.4 million US dollars was received by the country as foreign aid. CSR is not mainstreamed in the nation's business practice, corporate governance and corporate responsibility. As Porto (2010, p. 350) notes, "the principal arena for discussing CSR in Sierra Leone is in academic circles … governments have been characterised by corruption, mismanagement and inefficiency, only perpetuating the poor governance practices of their predecessors".

The preceding insights make Sierra Leone a resource-dependency nation and tormented by "resource curse" (Auty, 1993), a situation where a country is plagued by its natural resources such as diamond and bauxite. Resource curse theory (Karl, 1997), which is also referred to as "paradox of plenty" is central to 11 years of civil war in Sierra Leone as well as other governance and leadership issues in the country. According to Christian Aid (2013), Sierra Leone is at the moment struggling to rebuild after over a decade of civil war that triggered "blood diamonds", which are "rooted in inequitable distribution and irrational exploitation of resources. … the country still ranks near the bottom of the United Nation's Human Development Index, with high levels of poverty and food insecurity" (p. 5). Thus, Sierra Leone is still in the shadow of underdevelopment.

Also, since 2009, it has been indicated that approximately more than one-fifth of Sierra Leone's arable land has been leased to mostly MNCs for industrial-scale agriculture and gold/diamond mining. Whilst land deals slightly impact positively on economic survival of the local communities, most CSR initiatives are not strengthening the economic capacity and autonomy of affected communities (Maconachie, 2016). Whilst attempts are made by MNCs in Sierra Leone such as Addax Bioenergy, Socfin Agricultural Company Ltd. (SL), Sierra Rutile Ltd., Koidu Holdings and Goldtree amongst others to be socially and environmentally responsible, stakeholders have routinely accused them of not genuinely engaging in CSR, as these CSR initiatives do not compensate for social and environmental challenges Sierra Leoneans face as a consequence of their operation in the country (Maconachie, 2008; Datzberger, 2014). Thus, this chapter maintains that CSR should be used to address social, economic and environmental challenges people face for sustainable development (Newell, 2005; Idemudia, 2010).

Defining CSR

CSR can be defined as an organisational business model of self-regulation, which deals with organisational responsibility, accountability, transparency and ethics that are enshrined in business practice beyond what the law requires (Crowther & Aras, 2008; Carroll, 1979). CSR contrasts with strategic organisational gains given its inseparable link with normative business model (Garriga & Mele, 2004). Thus, CSR is desirable, not required, and companies benefit a lot from engaging in CSR for social licence and legitimacy to operate and to justify their actions to stakeholders (Clarkson, 1995; Freeman, 1984). This perspective aligns with Johnson, Scholes & Whittington's definition, which indicates that CSR is premised on "… ways in which an organisation exceeds the minimum obligation" to its varied stakeholders specified via regulations and corporate governance requirements. Although there is no consensus on what CSR really means, which makes it "a contested concept" (Okoye, 2009), it is basically an organisational philosophy and practice to make a profit but sensitive to social and environmental needs of society. For Idemudia (2010), this organisational practice is synonymous with "business *in* society" debate as opposed to "business *and* society". The former resonates with business helping in providing social goods; whilst the latter is about profit orientation of business (Idemudia, 2010).

Given the above, no matter the apparent lack of agreement on the meaning of CSR, there is consensus on what it means. It is about giving back to society in contemporary business era, an epoch marked by increasing stakeholder pressures as well as other externalities (Amaeshi, 2010) for companies to move from just thinking about the interest of shareholders to considering interest of anyone who can affect or be affected by its operation (Freeman, 1984). In furthering this position, following increasing social and environmental expectations of businesses in advancing community development (Blowfield & Murray, 2011), alleviating poverty

(Idemudia & Ite, 2006), helping in building local capacity (Frynas, 2005; Boele et al., 2001) propelling social investing (Amaeshi, 2010) and filling governance gap (Scherer & Palazzo, 2011), CSR is conceived to place premium on "business in society" thesis. Legitimacy and organisational reputation are two vital intangible resources accruable from CSR that can make MNCs in Sierra Leone have social licence to operate smoothly and harmoniously with their various stakeholders. Thus, organisations that do not pay attention to pressures from stakeholders stand the risk of undermining sustainable development (Emeseh, 2009) as well as risk corporate–stakeholder conflict (Katsos & Forrer, 2014) and credibility gap (Deegan, 2002).

Origin of CSR

CSR is generally conceived as an Anglo-American invention. Following the Joint Stock Companies Act (1844) in Britain as well as the parliament passing the Limited Liability Act in 1855, efforts were made towards reconceptualising the role of business in society. Across the Atlantic—in the United States—the US Supreme Court in Santa Clara vs. Southern Pacific Railroad ruled that companies should be seen as persons. Thus, by giving companies "corporate veil", they should also be conceived as persons in relation to being responsible for their actions as well as their impacts on society. This singular action framed largely the creation of the Fourteenth Amendment that empowered companies to have privacy rights with accompanying right to be taken to court, should they violate laws.

The initial writing on the concept "can be traced to the exchange of articles in the 1930s between Berle and Dodd on the role of corporate managers" (Okoye, 2009, p. 613). According to writers (Carroll & Shabana, 2010; Boulding, 1956; Bowen, 1953; Kreps, 1940; Clark, 1939; Bernard, 1938; McGuire, 1963; Thompson, 1967; Garriga & Mele, 2004), the history of CSR can be traced to the 1950s with the writings of Bowen (1953), who is generally considered as "the father of CSR". In the 1950 CSR was referred to as *social responsibility,* not *CSR*. The reason for this can be explained from the perspective that corporations at the time were not considered as part of society. In sum, these writers stressed that companies should give their activities in society a human face. Also, following Friedman's (1970) work urging businesses to increase shareholder value, the Committee for Economic Development came up with a seminal report that helped in redrawing the business–society relationship. To this end, a renewed CSR blueprint emerged in the 1970s replacing what Wood (2010) called "muddled and ambiguous" (p. 52) interpretation of CSR, which has taken the CSR debate further in relation to normative approach to CSR.

Following from the foregoing, Frederick (1994) identified what he called CSR1, which follows the trend of the early conceptualisation of CSR. Although this type of CSR stresses that businesses have an obligation to work for social benefit (Carroll & Shabana, 2010; Wood, 2010), this may be a legal or voluntary obligation, where profit is the focus. Opponents of this CSR type claim that the obligation is severely limited by financial, economic and profit considerations (Freeman, 1984). In CSR1, the company follows the preferences of its management, which suggests that CSR1 prioritises the firm's "enlightened self-interests" over others', in order to gain corporate sustainability. Frederick (1986) supposes that CSR2 (Corporate Social Responsiveness), refers to the firm's ability to respond to social issues and pressures. Unlike the CSR1, which is steeped in philosophical rationality, CSR2 adopts a managerial method. CSR2 emphasises the management of an organisation's relations with society (Carroll, 1979) as well as focuses on the long-run role of the firm in a dynamic social system (Sethi, 2007).

Subsequently, Frederick (1986) made an effort to fill the "values" gap in CSR1 and CSR2 by suggesting CSR3, which he regarded as "Corporate Social Rectitude". This model of CSR incorporates moral and ethical dimensions to CSR debate (Frederick, 1986, p. 134). He also categorises three value-based and/or ethical motivations for managerial decision-making including human rights (individual concerns of non-manager stakeholders), utilitarian (economic self-interest of the company) and social justice (distributional or societal-level concerns) (Mitnick, 2000). Nevertheless, Frederick's (1986) CSR3 has been subjected to criticism by scholars arguing that its "overzealous ethics push might reduce business productivity" and might "hamper society's vital economising process" (Frederick, 1986, p. 8). Other criticism includes the subjective nature of "values" which may differ across cultures, settings, groups and cultures. Continuing his normative model and borrowing insights from Mitnick's (2000) "normative referencing" notion, Frederick proposed CSR4 (Corporate Social Reason), which deals with issues of the Cosmos, Science and Religion. Everything, including business operation, is a consequence of cosmological mechanisms and processes, and to understand the Cosmos, we need to turn to Science, by going beyond social and behavioural sciences and putting into consideration Third Culture Science, which is a normative science. Thus, companies base their decisions on the CSR4 consciousness, which revolves around the Cosmos, rather than the company. In this framing, companies ought to take into consideration issues around sustainable development such as the environment and social issues. This proposition is central to normative culture and sustainable relationships, which resonates with the ideals of sustainable development (KPMG, 2017).

However, the leading theory on CSR in the 1970s was Carroll's (1979) work known as "CSR Ladder" and subsequently "CSR Pyramid" (Carroll, 1991), which analysis undertaken here is based. These two theorisations of CSR laid the foundation for contemporary conceptualisation of the notion (Matten & Crane, 2005). Carroll's (1979) "CSR Ladder" is graphically represented in Fig. 9.1.

In extant literature, Carroll's (1979) model is the most cited CSR model (Frynas, 2009; Matten & Crane, 2005). Carroll's (1991) "CSR pyramid" corresponds with views shared by various scholars (Idemudia, 2010; Frynas, 2009; Ite, 2004; Idemudia & Ite, 2006) for business to be part of society. CSR pyramid is graphically illustrated in Fig. 9.2.

As indicated by Fig. 9.2, for CSR to be conscientious, it should be framed in such a manner that the entire range of business responsibilities is embraced (Carroll, 1991). Carroll (1991) suggests that four types of social responsibilities constitute

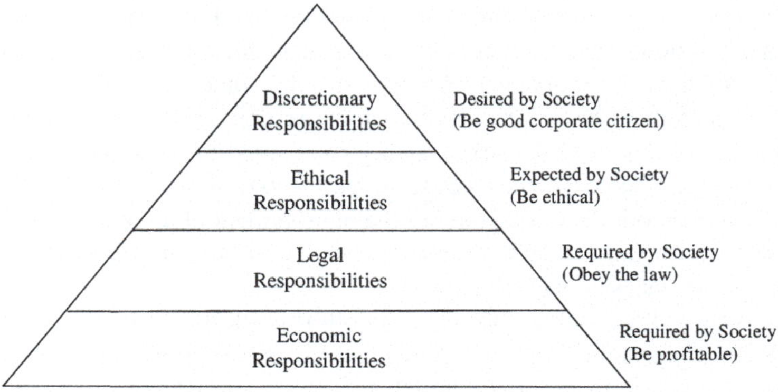

Fig. 9.1 CSR ladder. Source: Carroll (1979, p. 499)

Fig. 9.2 CSR Pyramid. Source: Carroll (1991)

total CSR including economic, legal, ethical and philanthropic. Moreover, these four elements or components of CSR can be represented as a pyramid. The pyramid of CSR depicts the four elements of CSR, starting with the basic building block view that economic performance undergirds all else (Visser, 2006). Also, organisations are expected to obey the law because the law is society's codification of acceptable and unacceptable behaviour (Okoye, 2009). Next element is organisation's ethical responsibility. Simply, this is the obligation for businesses to do what is just, fair and right, and to reduce or avoid harm to stakeholders including consumers, employees, the environment, communities, amongst others. Finally, organisations are expected to be a good corporate citizen (Eweje, 2007). This is considered in the philanthropic responsibility, wherein businesses are expected to contribute financially and socially to the community and to improve people's quality of life and the environment (Crowther & Aras, 2008). In contrast to just making a profit for organisations, managers are urged to be more sensitive to the interests and aspirations of other stakeholders whose interests can impact or can be impacted by their actions (Freeman, 1984; Clarkson, 1995). These contrasting viewpoints make CSR an "essentially contested concept" (Miles, 2012).

CSR as an "essentially contested concept"

CSR is a concept with definitional problems as it means different things to different groups or people (Frynas, 2009; Crane et al., 2008; Garriga & Mele, 2004, p. 51). It is considered as an "essentially contested concept" (Miles, 2012). Essentially contested concept is a term, whose precise definition is controversial and imprecise. Thus, the meaning of CSR is unclear (Okoye, 2009). Gond & Moon (2011, p. 4) assert that the concept is rather chameleonic—it changes its form to align itself to situations in the context of management ideals and orientation. Lack of consensus on what CSR means has become a critical factor in apprehending its conceptual dimensions, application and theory. This is because theoretical precision aids in enabling and sustaining development of a concept in a manner that facilitates collective understanding by creating "common frame of reference" (Miles, 2012, p. 285). This process can help in making such a concept less complex to understand. The difficulty in understanding CSR makes it problematic to measure the role of business in society (Siltaoja & Onkila, 2013). However, as noted by writers (see Siltaoja & Onkila, 2013; Idemudia, 2010; Crane et al., 2008) CSR is about giving back to society in order to have a more sustainable relationship with wider stakeholders.

Theorising CSR

The growth and theoretical development of CSR have facilitated researches aimed at theoretical approaches to explain the phenomenon (Frynas, 2009). As noted earlier, the pioneering work in this direction is Carroll's (1979) "CSR Ladder". CSR is in constant theoretical evolution, reinvention and development. Thus,

> There is no accepted theoretical perspective or research methodology for making sense of CSR activities. Indeed, most scholars study CSR without any reference to a given theoretical perspective, and it has been found that CSR research is not driven by continuing scientific engagement but by agendas in the business environment (Frynas, 2009, p. 12).

In considering the above perspective, in the management literature there are different approaches to theorise CSR. Frynas (2009) has indicated that business–society interface can be theorised from the following perspectives: "agency theory, stakeholder theory, stewardship theory, institutional theory, game theory, theory of the firm and resource-based view in strategic management" (p. 13). Despite lack of consensus on developing a coherent body of theoretical approaches to CSR (Belal, 2008), there are dominant theories on the concept. One of the often-cited theoretical positions is developed by Lantos, who proposes three theories of CSR: *altruistic*, *ethical* and *strategic* approaches. Altruistic CSR is pursued on the basis that there is no economic benefit to CSR. Strategic CSR is pursued by companies for reputation effect and cost minimisation; whilst ethical CSR entails morality of business (Lantos, 2001).

Carroll and Schwartz (2003) developed "Three-Domain Model of CSR". This framework helps in outlining three factors including economic, ethical and legal obligations of firms by collapsing the fourth domain—philanthropic or discretionary obligation—that was previously theorised by Carroll (1979, 1991). This theory is represented in Fig. 9.3.

In Carroll and Schwartz's (2003) view, the benevolence (philanthropic) responsibility element is omitted, which suggests that philanthropy is discretionary. Nevertheless, for moral and ethical organisational operation, this element is integral. Further to this, researches by Kang and Moon (2012) and Mitnick (2000) establish that the underlying concept of CSR is based on making profit and obeying the law as well as companies being ethical and good corporate citizens.

Further to the above, Garriga and Mele (2004) theorised that CSR can be categorised into four parts. The first typology is *instrumental theories*, which emphasise that CSR is fundamentally a way of arriving at profit maximisation for shareholders (stockholders) (Jones, 1995). This instrumental perspective is echoed here:

> Although some firms have committed to investments in CSR through the allocation of more resources, other companies have resisted. This could, at least in part, be because of the debate on whether a corporation should go beyond maximizing the profit of its owners as the only social responsibility of business, to being accountable for any of its actions that affect the people, communities and environments in which they operate (Samy, Odemilin & Bampton, 2010, p. 204).

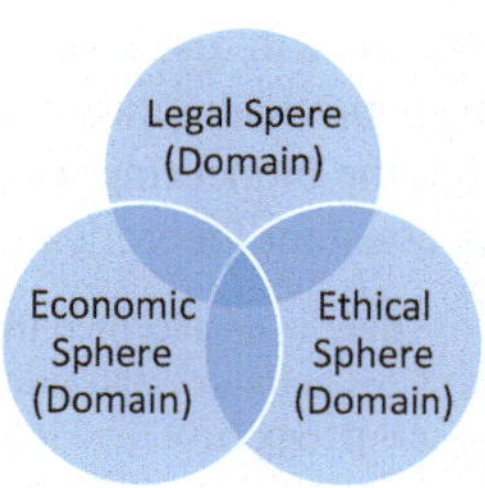

Fig. 9.3 Three-domain model of CSR. Source: Carroll and Schwartz (2003)

Some authors have argued that rationalising CSR along this axis potentially celebrates and promotes economic interests of shareholders at the expense of wider stakeholders (Idemudia, 2014; Visser, 2013). *Political theories* of CSR as theorised by Garriga and Mele (2004) concentrate on social powers of corporations acquired by using CSR to gain social advantage. Lately, there is an emerging body of research that considers politicisation of organisational practice (Matten & Crane, 2005; Whelan, 2012; Scherer & Palazzo, 2011). The third typology—*integrative theories*—considers theories based on the notion that businesses depend on society to succeed as well as their continued social legitimacy and existence (Idemudia, 2014). Proponents of this approach (Idemudia, 2010; Kleine & Hauff, 2009) contend that CSR as a business stratagem should be integrated into organisational strategy. The fourth, *ethical theories,* are hinged "on the principles that express the right thing to do or the necessity to achieve a good society" (Garriga & Mele, 2004, p. 60), are the hallmark of CSR.

In criticising Garriga and Mele's (2004) position, Secchi (2007) developed other ways of theorising by stating that Garriga and Mele's (2004) approach is essentially ambiguous. As he contends, political and instrumental approaches to CSR are just changes of name. Secchi (2007) proposes developing a CSR theory based on identifying where organisational responsibility is apportioned. This approach is expressed in *managerial*, *relational* and *utilitarian* approaches to CSR. *Managerial* approaches are essentially based on internal CSR drivers; whilst the *utilitarian* approaches consider CSR from the perspective of broader mechanisms of firms' self-interest and marketised system as main driving forces (Renouard, 2011). The *relational* approaches consider business–society interface as basically interpenetrating. In a similar vein, Windsor (2006) proposes three approaches including *corporate citizenship*, *instrumental* and *political* typologies.

In furthering the development of CSR, Dahlsrud (2008) proposes five typologies: *economic*, *environmental*, *social*, *stakeholder* and *voluntariness*. In addition, Lockett et al. (2006) offered four approaches: *environmental*, *ethical*, *social* and *stakeholder*. Nevertheless, regardless of theoretical approaches to CSR, an issue that stands out is that firms have to ensure corporate social performance (CSP) through CSR. It is to this end that Wood (2010, p. 693) defines CSR as an organisational practice and philosophy of social responsiveness as it interacts with its various constituents to perform as seen in Fig. 9.4.

Fig. 9.4 Corporate Social Performance Model. Source: Redrawn from Wood (2010)

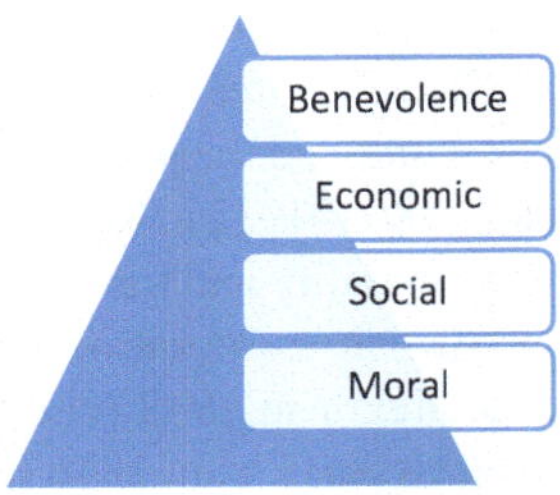

Conceiving CSR in this way aligns with what Orlitzky (2008) considers being wider and deeper explanation of the concept, which takes into account three norms at different levels of analysis: "*institutional*, *organisational* and *individual*" (Orlitzky, 2008, p. 115). Other theories of CSR have stressed economic factors (see Schaltegger & Wagner, 2006; Orlitzky et al., 2003), institutional factors (Campbell, 2000; Frynas, 2009) and legitimacy issues (Deegan, 2002; Castelló & Lozano, 2011).

Organisational Philosophy and Sustainable Development Debate: A Focus on why MNCs Engage in CSR in Sierra Leone

Although MNCs' underlying organisational and ideological philosophy is based on fierce capitalist pursuit (Bakan, 2004), the 1980s marked a watershed in international business venturing precipitated by economic liberalisation, globalisation and common global market that are unremittingly profitable to MNCs to outsource raw materials, labour and manufacturing goods in developing countries such as Sierra Leone (Maconachie, 2012). Developing countries' environment is quite attractive because of cheap labour supplies, poor labour regulations, lower wage payment system and unregulated CSR practices, which undermine sustainable development (Idemudia, 2010). MNCs' business operation globally—especially in developing countries—has triggered CSR issues (Amaeshi et al., 2006), political conflict (Maconachie, 2016), stakeholder conflict (Datzberger, 2014), wars (Binns & Maconachie, 2005), crushing poverty (Maconachie et al., 2006) and sustainability issues (Emeseh, 2009).

The above contention makes CSR an important issue in today's business venturing (Obi, 2010; Amaeshi et al., 2006). Additionally, CSR has become an important issue following grand corporate failures, business irresponsibility and poor accountability issues, which undermines sustainable development (Emeseh, 2009; Visser, 2013). This situation has been characterised by Silberhon & Warren (2007, p. 352) as "false dawns in past decades" regarding sustainable, ethical CSR practice. The false dawn resonates with corporate sagas such as Enron, Parmalat, Niger Delta conundrum, WorldCom, climate change, global warming and civil war in Sierra Leone over diamond mining and unsustainable agricultural practice by MNCs (Maconachie, 2016).

Sierra Leone is a country endowed with huge deposits of natural resources but has witnessed the presence of MNCs in its territory dated back in the era of colonialism (1808–1912) (Konteh, 1997). Wilson (2015) observed the lack of corporate responsibility and transparency by the then-colonial governor about the engagement of British corporations in the exploitation of the country's forest resources. MacGregor in 1942 (cited in Konteh, 1997) details the illegal operation of British firms as early as in 1816 (around the Bullom Shores, Great Scarcies and Port Loko) by a well-known British entrepreneur called John McCormack (Wilson, 2015). In addition, the influence of about a decade of civil war and crisis as well as poor governance exacerbated a situation of intense exploitation of Sierra Leone's natural resources by MNCs and poverty (Smillie et al., 2000). The increasing levels of poverty in the country made it possible for businesses to exploit the people and their environment including poor regulatory regime (Maconachie, 2012).

Given this situation, it is difficult for MNCs to support CSR initiatives that are meant to help communities grow at a sustainable level as well as invest in local development initiatives (for example, road construction and artisan centres for the locals). There are evidences supporting poor CSR practice of MNCs in developing countries like Sierra Leone (Wilson, 2015; Idemudia, 2010, 2014). Research on CSR suggests that corporate organisations can be instrumental in addressing sustainable development issues around the world (Emeseh, 2009). Theorists have considered why companies engage in CSR from the angles of legitimacy (Deegan, 2002), organisational image (Hooghiemstra, 2000), return on investment (ROI) (Ioannou & Serafeim, 2014) and public relations (Idemudia, 2010) amongst others. Some of the main reasons why organisations engage in CSR are highlighted below.

CSR and Organisational Legitimacy

According to Suchman (1995) organisational legitimacy is a function of how society approves of an organisation's conduct and activities in society, which resonates with "social licence" to operate (Idemudia, 2010). MNCs, in particular, constantly change the dynamics of their organisational practice to suit situations so as to have legitimacy (Dowling & Pfeffer, 1975). Organisational legitimacy can be strategic or normative (Siltaoja, 2009; Campbell, 2000). It is strategic when an organisation's managers manage stakeholders' expectations and interests in a way that the management (shareholders) feel that managers are acting in their best interests. When their actions are directed towards external stakeholders for social approval, it is referred to as normative legitimacy (DiMaggio & Powell, 1983).

The above contention is based on the assumption that organisations are embedded within the environment of a super-ordinate system, within which, society approves of their conduct as long as their actions are in agreement with wider societal norms and values (Parson, 1960). In acting within the bounds of a wider set of values and expectations, which Donaldson and Dunfee (1994) call "hyper-norms", organisations are considered wider stakeholders as acting ethically, normatively and responsibly in relation to CSR (Belal, 2008). Legitimacy theory has been studied variously by different authors (Belal & Owen, 2015; Brown & Deegan, 1999; Guthrie & Parker, 1989; Deegan, 2002). For instance, using legitimacy theory, Brown and Deegan (1999) maintain that media attention reflects community interests and aspirations. In a similar vein, in adopting legitimacy theory Belal and Owen

(2015) empirically concluded that by using stand-alone CSR reporting, MNCs in Bangladesh legitimise their operation in a country, where "both tobacco control regulation and a strong anti-tobacco movement were gaining momentum" (p. 1160).

CSR and Reputation Management

Theorising reputation and organisational image take the form of instrumental management approach that is framed by the contention that reputation is a social resource. Sociologically shaped standpoint to this understanding sees reputation as the outcome of shared socially constructed impression of organisations (Bebbington et al., 2008). In the view of Fombrun and Van Riel (1997), reputation constitutes subjective collective assessment of the reliability and honesty of organisation as it is second-order and derivative characteristic that arises from a specific organisational life. Thus, it can be argued that reputation management is a function of organisational actions aimed at swaying stakeholders' opinions about normative business practices. It is also concerned with efforts targeted at managing stakeholders' expectations, feelings and views about how a company is operating in a specific environment such as Sierra Leone. Managing reputation and image helps to reduce social risk (Bebbington et al., 2008; Hooghiemstra, 2000).

CSR Organisational Performance and ROI

In extant literature, CSR helps to increase ROI. In a study by Ullman (1985), at the heart of CSR is financial performance. Empirical researches considering the interface between CSR and financial advantage (ROI) take two main approaches. First, some of the studies demonstrate this by applying event study approach to methodologically explain that there are short-term financial effects when firms do not engage in CSR (Margolis & Walsh, 2003). This business method is expressed in the study by Wright and Ferris (1997), where they demonstrated a negative relationship between CSR and firm performance. In the study by Posnikoff (1997), it is noted that a positive relationship exists between CSR and financial gain. Second aspect of studies concentrates on locating the relationship between financial gain and CSP model by using financial indices of competitive advantage and profitability. Although opinions are polarised on this argument, however, by adopting CSR and accounting-based performance methods Cochran and Wood (1984) concluded that there exists a positive correlation between CSR and accounting performance. In echoing this, Waddock and Graves (1997) found appreciable level of positive association between an index of CSP and performance variables. Thus, there is a lot that organisations can benefit from engaging in CSR from a profit-oriented perspective.

From Sustainability to Sustainable Development

To fully understand the emergence of sustainability into mainstream literature, it is important to investigate the wider roots from which the concept emerged (Purvis et al., 2018). This argument is confounded by the fact that much of the scholarship and research that feed into the notion of sustainability predate the language of sustainability (Purvis et al., 2018). Writers have shed light on the rise of sustainability debate (Purvis et al., 2018; Grober, 2012). One of the early developments was insights from forestry experts of the seventeenth and eighteenth centuries who introduced the term "sustainable yield" as a response to diminishing forestry resources across Europe (Purvis et al., 2018). The efforts of political economists in the wake of industrial revolution questioning the limits of demographic and economic growth as well as the inherent trade-offs between social, equity and wealth creation, were also significant in the rise of sustainability concept (Lumley & Armstrong, 2004).

Additionally, the activities of natural scientists of the nineteenth and twentieth centuries culminating in a clarion call to understand the difference between anthropocentric conservationism and bio-centric preservationist discourse. They prescribed conserving the natural resources for sustainable consumption. The modern language of sustainability was popularised in 1972 (The Ecologist, 1972) with the creation of a "sustainable society" injunction. For brevity and given the remit of this study, the 1972 UN Conference in Stockholm marked the first global summit to address the impacts of human activities on the environment as well as first major effort to reconcile (economic) development with environmental concerns and integrity that were generally considered as incompatible (Purvis et al., 2018). Following this summit was the notion of "environmentally sound development", which was later termed "economic development" (Mebratu, 1998).

Sustainable Development: Genesis, Meaning and Implications

Sustainability having been arguably subdued in the 1980s the twin social and ecological appraisal of economic development started to interlace with economic development in relation to what has been termed "sustainable development" (World Commission on Environment and Development [WECD], 1987; Barbier, 1987). Furthermore, sustainable development discourse came to the fore on the premise of rethinking economic growth and social and environmental profit accruable when businesses deal with social issues including when companies (MNCs) give back to society by way of CSR resonating with social justice and community development. For more than 30 decades, the starting point for debate about sustainable development was the publication of *Our Common Future*, which is also referred to as the Brundtland Report by WCED (WCED, 1987; Crowther & Aras, 2008). Also, at the 2005 World Summit in New York following the United Nations' 2000 Millennium Summit, a forerunner to the MDGs, the issue of sustainable development was given a central place in order to address one of mankind's greatest challenges in the new century.

The concept of sustainable development is currently appearing in a great number of discourses and debates (Purvis et al., 2018). Nevertheless, like CSR, sustainable development integrates different issues, concepts and terms including sustainability, development and capacity building as well as related phenomena, which are generally used interchangeably (Ihlen & Roper, 2011). The notion is drawn from a range of aspirations and issues relating to development, peace, freedom, capacity and the environment (Purvis et al., 2018). Contemporarily, sustainable development as a concept, movement, goal or thought is fundamental to the objective of numerous corporate enterprises, businesses, national institutions and sustainable cities (Emeseh, 2009). The Brundtland Commission therefore suggested the current classical definition of the concept: "development that meets the needs the needs of the present without compromising the ability of future generations to meet their own needs" (WCED, 1987, p. 16).

For our purpose in this study, sustainable development means a participatory process that generates and pursues a vision of the community, which respects as well as makes prudent use of its resources—human, natural, human-created, cultural, social and other means—for future co-existence and conflict-free relationship (Filho & Brandli, 2016; Hemmati, 2002). As noted by Desjardin (2000) sustainable development (sustainability) has grown out of the recognition that economic development on a global level cannot be separated from issues regarding social justice and ecological stability. Thus, "The new worldview emerging as an alternative to the reigning paradigm of economic growth and free markets holds that long-term sustainability is the criterion of successful economic and social development" (Desjardin, 2000, p. 10–11). This process recognises that doing business should not jeopardise the wellbeing of the environment and the people both now and in the future (Bebbington, 2000). This process parallels Visser's (2013) contention that sustainable development attempts to ensure, to the degree possible, that present generations reach a high dimension of economic prosperity and safety, whilst maintaining the integrity of the ecological systems on which all life and all production is based (Visser, 2013).

Nevertheless, some writers (Chichilinisky, 1994; Parris & Kates, 2003) have noted that sustainable development is ambiguous as it could mean different things to different people. For instance, Greenpeace has described the concept as mere "deceptive jargon" of anti-environmentalism (Hayward, Fowler & Steadman, 2000, p. 45). Despite the criticism, central to its conceptualisation are development, social participation and protection of the environment (Purvis et al., 2018). Nonetheless, the notion of sustainable development is being adopted to deal with a range of developmental challenges (Crowther & Aras; Hayward et al., 2000). Due to further ambiguity surrounding the meaning of the concept of sustainable development, other definitions are based on "what it specifically seeks to achieve" and "how it is measured", for example indicators including the Environmental Sustainability Index, the Global Reporting Initiative and Wellbeing Index amongst others (Parris & Kates, 2003).

Sustainable Development and Businesses

Despite the ambiguity of the concept—sustainable development—it is widely adopted in business strategy and operation for a more harmonious and sustainable relationship between businesses and society as well as other stakeholders (Conteh & Maconachie, 2019). According to the World Business Council for Sustainable Development (WBCSD, 1987) sustainable development is:

> The form of progress of business strategies and activities that meet the needs of the enterprise and its stakeholders at present while protecting, sustaining and enhancing the human and natural resources that will be needed in the future (WBCSD, 1987, p. 1).

The above principle has been taken into consideration in the "Maastricht and Amsterdam Treaties on European Union, as well as in the Rio Declaration and Agenda 21", which is "adopted by the United Nations Conference on Environment (UNCED), meeting in Rio de Janeiro" in 1992 (Crowther & Aras, 2008, p. 41). The above definition rekindles interest in the original conceptualisation of the notion in the Brundtland Report as well as identifies that economic development must meet the needs of businesses and their numerous stakeholders.

There are seven main issues identified in the Brundtland Report for sustainable development:

- Changing the quality of growth.
- Ensuring a sustainable level of population.
- Reorientation technology and managing risk.
- Reviving growth.
- Managing environment and economics in decision-making.
- Meeting essential needs for job, energy, food, water and sanitation.
- Conserving and enhancing the resource base.

The Brundtland Report makes legal and institutional recommendations for change so as to deal with common global issues (Crowther & Aras, 2008). The above definition also brings to attention that businesses depend on natural and human resources that should be preserved for future generations (WBCSD, 1987). From this viewpoint, organisations including businesses (micro-level) are called upon to address social, developmental, environmental and economic issues at the macro-level (societal level) including social equity (human rights, community and poverty); environmental accountability (biodiversity, climate change, global warming and land use); and economic efficiency (prosperity, innovation, productivity). According to Pan et al. (2018) and Loew et al. (2004) micro and macro-level issues are different but interlinked. This perspective further encourages businesses and their managers to consider sustainable development in regards to business operation, strategy and engagement for sustainable development (Filho & Brandli, 2016).

The Business Approach to Sustainable Development

According to Blowfield and Murray (2011) the business approach to sustainable development implies fair distribution across the world's population in relation to quality of life (WBCSD, 1987). As they further noted, this has to be realised by businesses without hindering the ability of future generations to meet their own needs in the context of eco-efficiency, eco-justice and social justice and investing. Achieving this goal is central to tackling poverty, climate change issues and social inequality that are rife in developing countries as opposed to developed countries—particularly sub-Saharan Africa, which is the context of this study—Sierra Leone. WBCSD (1987) has continually stressed that issues such as climate change, global warming, poverty, inequality, good global governance and education are responsibilities of the whole world including businesses.

As noted by Elkington (1997), these challenges present businesses with a dilemma about how to adapt inclusive and sustainable business practices to organise profit-making activities to deal with the challenges highlighted above, whilst maximising profit. Elkington (1997) further observed that such sustainable, inclusive business model will advance the win-win approach (Williams & Preston, 2018) for both companies and society. Furthermore, modelling business practices along the lines of sustainable, inclusive stakeholder relationship brings to light the general consensus that sustainable development sustainability issue challenges all and sundry to understand the social, environmental and economic aspects of sustainability in an integrated way. Indeed, the Brundtland Report highlighted three key elements or dimensions of sustainable development, which are:

- Social equity
- Environmental protection
- Economic growth

These three "pillars" of sustainable development (sustainability) have been commonly recognised and adopted as the most significant characteristics of sustainable development (see Purvis et al., 2018; Crowther et al., 2019; Crowther & Aras, 2008; Elkington, 1997). The three interconnected "pillars" conception of sustainable development are "the dominant interpretation within the literature" (Purvis et al., 2018, p. 1). They parallel Elkington's (1997) three pillars of sustainability: planet (environment), people (equity) and profit (economy) for sustainable development. Taken together, CSR is considered in this research as a guiding business model, which addresses both short- and long-term social, environmental and economic performances of a business. On the other hand, sustainable development emphasises three pillars in the context of long-term wellbeing (Crowther et al., 2019). Therefore, CSR calls on businesses all over the world to address issues identified in their environment and communities beyond their main objectives (Belal, 2008).

MNCs and CSR in Developing Countries

As noted by Bakan (2004) MNCs are institutions with great power that is wielded over (developing) nations and society. It has been contended that given their ease of mobility, capital base and power as well as apparent lack of loyalty to a given jurisdiction, MNCs are traditionally mistrusted by countries, where they are found (Zerk, 2006). In Europe, the controversy surrounding the MNCs is a post-war event, when expansion was expedient to resuscitate failing (or failed) corporations because of World War II. Also, the expansion and power of MNCs were propelled by public (local authority) resentment regarding overreliance on FDI from the USA. This situation also includes the "Americanisation" of management, culture and lifestyle. On the other hand, in the USA, MNCs appear to have been considered relatively benign by the public and government until the 1960s, when their reputation took a decline resulting in public criticism of their actions in society (Frederick, 1994; Carroll, 1999). By the 1970s, the MNCs have become synonymous with global influence, power, wealth and dominance. Extant literature on this phenomenon indicates the degree to which corporations are associated with interest of their state of origin—home countries. They thus depend on less developed (or developing) countries to oil the wheel of their economic interests and activities. As Barnet and Muller (1974) stated, MNCs are conceived simplistically as economic and political agents of their home countries, with no specific allegiance to the nations where they do business or where they invest.

Therefore, the relationship between the MNCs and "host countries" has not been cordial to say the least as their host nations see them as jeopardising their economic prosperity and self-reliance, hence, reason for CSR controversy (Frynas, 2009; Maconachie, 2016). In addition, foreign-owned MNCs are deemed to be a threat to the sovereignty and political leadership of their host countries given the fact that they might exercise undue power and influence on these countries as well as aid deepening of inequality and underdevelopment. These reasons and more prompted the UN's Economic and Social Council (ECOSOC) to launch a research project aimed at addressing the activities of the MNCs and their effects on the process of development. Issues such as trade liberalisation, globalisation and the rise of fierce capitalism at the turn of the millennium accentuated and sustained the rise of the MNCs. However, their rise has come with severe criticism from stakeholders and their host countries (Visser, 2013).

According to Zerk (2006), the term multinational is a flexible and fluid concept that has evolved over time. Initially, the very notion of the concept meant any company or firm that is owned directly or via its subsidiaries' capital or assets located in the territory of more than one country or state (Fieldhouse, 1986). These firms are also run on an integrated basis, which enables smooth operation and transfer of capital from host countries to home states. However, the most distinguishing factor that makes MNCs different from other forms of investment or "portfolio" investment, is the presence of FDI. FDI is not the only thing that MNCs bring to the table in relation to operating in periphery nations: it also controls resources and power as well as transfers technology and management style, which preclude development in host countries. It is important to understand who or what is multinational or what it represents. As considered by UNCTAD (2000, p. 267), MNCs can be defined as incorporated or unincorporated firms consisting of parent companies and their foreign associates. A parent company can be interpreted as a firm that controls capital/assets of other entities in countries that are not its home country. This is operationalised by having equity capital stake. In the context of this study, a multinational can be defined as a parent enterprise located in a home state and connected with its foreign associates (affiliates) via affiliation of control—cross-border control nexus.

Nevertheless, MNCs are rather seen by wider stakeholders as not living up to their promises with regard to CSR for sustainable business enterprise (Cartwright, 2007; Dicken, 1998). The dominant motif in the extant literature is that although the presence of the MNCs has good sides to economic state of periphery nations such as Sierra Leone, yet, there is increasing pressure from wider stakeholders for more ethical business venturing. Thus, although the dynamics of globalisation and business internationalisation have propelled "vast opportunities for the multinationals from the global south—foreign direct investment outflow from developing countries having increased from mere 5.2% in 1990 to 14.2% in 2006—it remains to be seen whether MNCs with their poor record of CSR would modify their behaviour" (Raman & Lipschutz, 2010, pp. 5–6). Additionally, the presence of MNCs in developing countries has spawned criticisms and corporate–stakeholder deadlock as their organisational philosophy is in contrast with stakeholders' interests (Benito & Narula, 2007).

MNCS, CSR and Sustainable Development in Sierra Leone

MNCs, which are also called Transnational Enterprises (TNEs), Multinational Enterprises (MNEs) or worldwide enterprises are corporate organisations that own or control production of goods and/or services in at least one country other than their home country. Zerk (2006) stated that a firm or group of firms should be regarded as a multinational corporation if they derive 25% or more of their total revenue from out-of-home-country activities and operations. Although MNCs contribute to economic development in developing countries (Zerk, 2006), scholars (Chan & Ross, 2003; Tonelson, 2002) argue that they gain more by internationalising their operations. This explains the rationale for "the race to the bottom" (Tonelson, 2002). As organisations that benefit hugely from their operations in host countries through governments' grant of lower tax and labour costs to attract FDI as well as tax exemptions in certain instances, MNCs have an important role to play in sustainable development through CSR (Crowther & Aras, 2008).

Accordingly, Visser (2013) argues that MNCs can be instrumental in driving economic growth, reducing poverty and creating jobs, advancing prosperity and raising employment standards by paying higher wages than local businesses do. Additionally, they can contribute to national development by transferring knowledge, technology and advanced skills to build or improve social infrastructures, raise living standards and accelerate the pace of sustainable development. Although the presence of MNCs helps in sustainable development, their activities are questionable given weak institutional and governance regimes in developing countries (Zerk, 2006). Recent scholarship has emphasised the impact of institutional environment on practices of CSR by MNCs including industry-specific circumstances, organisation's home country pressure and their relationship with key stakeholders (DiMaggio & Powell, 1983).

These issues make contributions of MNCs to sustainable development less credible. This study also argues that CSR approaches and initiatives are embedded in and framed by wider social realities and institutions, for example corruption, political and governance systems and weak regulations (Amaeshi et al., 2006; Idemudia, 2010), which undermine normative CSR practice. Thus, institutional understanding of CSR practice by MNCs suggests that corporations do not fully comply with regulatory frameworks operative in countries they operate or that they do not legitimately engage in CSR. Specifically, Akiwumi (2014, p. 774) in his study on Sierra Leone contends that although MNCs make effort to abide by mining laws and regulations, "their operations contribute to the erosion of indigenous cultures ... exploit natural resources ... and violate the subsidiarity principle of sustainable development when they exclude communities from participating in development projects that affect them".

CSR is generally considered a corporate response to stakeholder claims (Belal, 2008). As firms with multiple stakeholders, the perspectives of key stakeholders including local communities are crucial to their survival and legitimacy (Freeman, 1984). Nevertheless, extant literature underscores the instrumental dimension of CSR as it relates to whether organisations including MNCs successfully engage in CSR. MNCs are naturally exposed to diverse stakeholders as they internationalise their operations as well as extensive stakeholder claims and criticism (Margolis & Walsh, 2003). Coordinated and well-executed CSR can be a critical strategic initiative for MNCs in host countries. Furthermore, CSR initiatives should reflect the values and goals enshrined in SDGs following the adoption of the 2030 Agenda for Sustainable Development. In the context of Sierra Leone, in a bid to salvage the ailing economy, the government has made effort to advance sustainable development ideals and CSR by introducing *Agenda for Change* as well as incorporating the principles of SDGs (Akiwumi, 2014), however, failure of government agencies to adequately monitor, for instance, forest reserve and mining sector, is attributable to the present challenges bordering on lack of sustainable development.

Increasingly, attention has been drawn to the destructive consequences of unsustainable practices by MNCs from governments, policymakers, researchers and international actors, regarding various facets of the activities of businesses to overcome the challenges of resources scarcities, mitigate environmental risks and to promote sustainable development (Crowther & Aras, 2008). Since CSR represents core values and aspirations that delineate the commitment of an organisation to economy, society and the environment that sustain its operations, its wider scope can be integrated into a sustainable development framework and agenda (KPMG,

2017). According to Visser (2013), it is not about the volume of economic initiative undertaken by a MNC, what is important is how such initiative positively impacts the people, economy and environment. Thus, CSR seems to further social causes and public goods beyond mere strategic, profit-making orientation of business (World Business Council for Sustainable Development [WBCSD], 1987; Crowther et al., 2019).

However, CSR is a subjective and debatable notion, because societal needs affect the norms, values and corporate principles attached to CSR practices, in a way that the notion of development has received disparate perspectives. The supply side argument proposes social responsibility of business to society; whilst the demand side advocates the increased desire for sustainable, safe and socially responsible contributions of business to society (Idemudia, 2010). These two opposing values are revealed in the contextualisation and conceptualisation of CSR (Idemudia, 2011); the ideal that managers should act in the best interest of shareholders, as agents contractually under the obligation to maximise wealth (Friedman, 1962); and lack of clarification on how firms should act in a socially responsible manner with regard to changing societal values, including increasingly complex business milieus (Margolis & Walsh, 2003).

Postscript

Despite the attempts above to ensure more accountability, transparency and responsible practice by MNCs, it has been variously reported that Sierra Leone faces one of the worst forms of CSR dilemma in Africa (Porto, 2010; Maconachie, 2008, 2009; Datzberger, 2014). Given the seemingly failure of various initiatives and agencies by Sierra Leone government such as the Kimberley Process Certification Scheme, Sierra Leone Extractive Industries Transparency (SLEITI) and The Diamond Area Community Development Fund (DACDF), this chapter contends that a critical exploration of the rationale for failure of these institutions and/or bodies will galvanise a better understanding of the way forward in Sierra Leone from the prism of sustainable development in Sierra Leone.

Although recently, literature on CSR in Africa has widened in theoretical scope and depth, it is still essentially in its incipient stage (Visser, 2006). For instance, Gokulsing (2011) and Ragodoo (2009) have widened scholarship on CSR in Africa by interrogating CSR practices in Mauritius; Kolk and Lenfant (2013) have also explored this phenomenon in the context of conflict in Central Africa. In a similar vein, Idemudia (2010) and Ite (2004) have all investigated CSR from the Nigerian context. In South Africa, some studies have been done to enrich scholarship on CSR (Visser, 2013). Consequently, Visser et al. (2006) argued that there is a need to increase the dimensions of CSR scholarship in Africa. Thus, CSR scholarship in Africa is a critical aspect of scholarly enquiry, "driven by the legacy of colonialism…, the human needs of the continent in the face of widespread poverty, and the trend towards improved social responsibility by multinationals in a globalising economy" (Visser, 2006, p. 18). This is also vital for sustainable development in Sierra Leone given the seemingly failure of recent efforts to galvanise better corporate governance and accountability in the country (Maconachie, 2008).

Therefore, in Amaeshi's (2010, p. 44) view, this is "where CSR complements the exiting public and informal government configurations, and thus creates a better chance that both the public and private government modes will compensate for each other's weakness in governance of corporate externalities". The idea of using CSR for sustainable development by MNCs is not a new practice in Africa (Idemudia, 2010). Noyoo (2010) demonstrated how an Anglo-American company as early as 1929 was (socially) responsible for "providing housing and hygiene for its employees in Zambia but with racial bias" (Qtd in Idemudia, 2014, p. 422). Similarly, Honke (2012) pointed out the nexus between the paternalistic dynamics of corporate-stakeholder intervention programmes engaged by mining firms in the colonial era and the post-colonial dispensation within the Democratic Republic of Congo (DRC). In Sierra Leone, FDI and similar transnational business enterprises have been established as rather a way of continuing the process of colonial hegemony, domination and exploitation in Africa and Sierra Leone in particular (Maconachie, 2008). Some scholars in Africa (Visser, 2006; Idemudia, 2010) have commented that CSR is merely a strategic instrument by companies to advance their profit-maximisation agenda including bettering their organisational reputation rather than a genuine commitment towards sustainable development and sustainability.

As noted by Crowther & Aras (2008, p. 41) "of the three principles of CSR" including transparency and accountability, sustainability is concerned with impacts that actions that humans take at the present time can affect the options available in the future (Purvis et al., 2018). This is why sustainability has become a common issue for business and society (Purvis et al., 2018). Although every business organisation mentions CSR and claims that it engages in CSR for sustainable development, however, this is not so as these claims can be mere rhetoric, green-washing or strategic tools by organisations. Nonetheless, although the relationship between business and society has been subjected to much debate, evidence still suggests that good and ethical organisations make a positive contribution to the environment and society (Siltaoja, 2009). Therefore, CSR should not just be used to further MNCs' economic and strategic interests; but a method of reconceptualising the role of business in society by factoring in Elkington's (1997) "triple bottom line" approach for sustainable development. This business model considers inseparable relationship between the economic, environmental and social in business venturing for sustainable development (Crowther & Aras, 2008). Accordingly, this much-desired CSR approach should be propelled not just by ideological and strategic rationality as well as the fact that companies can be powerful forces for social transformation, but by normative culture and responsible social investing that can advance prosperity on the African continent. This CSR philosophy is not only important for better corporate–stakeholder relations but for poverty reduction and/or eradication and sustainable development (Blowfield & Murray, 2011).

Conclusion

This chapter has interrogated the interface between CSR and sustainable development within the context of Sierra Leone. Central to the purpose here was to clarify understanding about the role CSR plays in sustainable development. Effort was made in this regard to define the concept as well as its origin and the contested nature of the phenomenon. Also, different theoretical as well as conceptual bases for explaining the concept are offered including the organisational philosophy argument as agent for sustainable development. There is also some elucidation about why MNCs engage in CSR, particularly in developing countries like Sierra Leone, argument about the transition from sustainability to sustainable development as well as the latter's genesis, meaning and implications. Additionally, the nexus between businesses applying the CSR model for sustainable development is delineated and the business approach to sustainable development. More so, the chapter presented how the MNCs in Sierra Leone could use the agency of CSR to facilitate and foster sustainable business practice that will not only enhance better relationship between firms and the diverse stakeholders but that will generate reputational capital, legitimacy and normative practice for lasting peace in the country.

References

Agbiboa, D. E. (2012). Between corruption and development: The political economy of state robbery in Nigeria. *Journal of Business Ethics, 108*(3), 325–345.

Akiwumi, F. (2014). Strangers and Sierra Leone mining: Cultural heritage and sustainable development challenges. *Journal of Cleaner Production, 84*, 773–782.

Amaeshi, K. (2010). Different markets for different folks: Exploring the challenges of mainstreaming responsible investment practices. *Journal of Business Ethics, 92*(1), 41–56.

Amaeshi, K., Adi, B., Ogbechie, C., & Amao, O. (2006). Corporate social responsibility in Nigeria: Western mimicry or indigenous practices? *ICCR Research Paper Series, 39*, 1–39.

Angus-Leppan, T., Metcalf, L., & Benn, S. (2010). Leadership styles and CSR practice: An examination of Sensemaking, institutional drivers and CSR leadership. *Journal of Business Ethics, 93*(2), 189–213.

Auty, R. M. (1993). *Sustaining development in mineral economies* (Vol. 19, pp. 14–29). Routledge.

Bakan, J. (2004). *The corporation: The pathological pursuit of profit and power*. Penguin.

Banerjee, S. B. (2008). Corporate social responsibility: The good, the bad and the ugly. *Critical Sociology, 34*(1), 51–79.

Barbier, E. B. (1987). The concept of sustainable economic development. *Environmental Conservation, 14*(2), 101–110.

Barnet, R. J., & Muller, R. E. (1974). *Global reach: The power of the multinational corporations*. Simon and Schuster.

Bebbington, J., Larrinaga, C., & Moneva, J. M. (2008). Corporate social reporting and reputation risk management. *Accounting, Auditing & Accountability Journal, 21*(3), 337–361.

Bebbington, J. (2000). Sustainable development: A review of the international development, business and accounting literature. *Accounting Forum, 25*(2), 128–157.

Bebbington, J., & Unerman, J. (2008). Editorial. *Social and Environmental Accountability Journal, 28*(1), 1–2.

Belal, A. R. (2008). *Corporate social responsibility reporting in developing countries: The case of Bangladesh*. Ashgate.

Belal, A., & Owen, D. L. (2015). The rise and fall of stand-alone social reporting in a multinational subsidiary in Bangladesh: A case study. *Accounting, Auditing and Accountability Journal, 28*(7), 1160–1192.

Benito, G., & Narula, R. (2007). *Multinationals on the periphery*. Ashgate.

Bernard, C. (1938). *The functions of the executives*. Harvard University Press.

Binns, T., & Maconachie, R. (2005). Reconstructing rural livelihoods in post-conflict Sierra Leone. In K. Jackson, N. Lewis, S. Adams, & M. Morten (Eds.), *Development on the edge* (pp. 100–106). University of Auckland.

Blowfield, M., & Murray, A. (2011). *Corporate responsibility*. Oxford University Press.

Boele, R., Fabig, H., & Wheeler, D. (2001). Shell, Nigeria and the Ogoni. A study in unsustainable development: I. The story of Shell, Nigeria and the Ogoni people—Environment, economy, and relationships: Conflict and prospects for resolution. *Sustainable Development, 9*(2), 74–86.

Boulding, K. (1956). General systems theory the skeleton of science. *Management Science, 2*(3), 197–208.

Bowen, H. R. (1953). *Social responsibilities of the businessman*. Harper.

Brown, N., & Deegan, C. (1999). The public disclosure of environmental performance information–A dual test of media agenda setting theory and legitimacy theory. *Accounting and Business Research, 29*(1), 21–41.

Campbell, D. J. (2000). Legitimacy theory or managerial reality construction. Corporate social disclosure in Marks and Spencer corporate reports, 1969-1997. *Accounting Forum, 24*(1), 80–100.

Carroll, A. B. (1979). A three-dimensional conceptual model of corporate social performance. *Academy of Management Review, 4*(4), 497–505.

Carroll, A. B. (1991, July–August). *The pyramid of corporate social responsibility: Toward the moral management of organisational stakeholders* (pp. 39–48). Business Horizons.

Carroll, A. B. (1999). Corporate social responsibility. *Business and Society, 38*(3), 268–295.

Carroll, A. B., & Shabana, K. M. (2010). The business case for corporate social responsibility: A review of concepts, research and practice. *International Journal of Management Review, 12*(1), 85–105.

Carroll, A. B., & Schwartz, M. S. (2003). Corporate social responsibility: a three-domain approach. *Business Ethics Quarterly, 13*(4), 503–530.

Cartwright, R. (2007). *Mastering the Globalisation of Business*. Palgrave.

Castelló, I., & Lozano, J. M. (2011). Searching for new forms of legitimacy through corporate responsibility rhetoric. *Journal of Business Ethics, 100*(1), 11–29.

Chan, A., & Ross, R. J. S. (2003). Racing to the bottom: International trade without a social clause. *Third World Quarterly, 24*(6), 1011–1028.

Chichilinisky, G. (1994). North–South trade and the global environment. *American Economic Review, 84*(4), 851–874.

Christian Aid. (2013). *Who is benefitting?* Christian Aid.

Clark, J. M. (1939). *Social control of business*. University of Chicago Press.

Clarkson, M. B. E. (1995). A stakeholder framework for analysing and evaluating corporate social performance. *Academy of Management Review, 20*(1), 92–117.

Cochran, P. L., & Wood, R. A. (1984). Corporate social responsibility and financial performance. *The Academy of Management Journal, 27*(1), 42–56.

Commission for Africa. (2005). *2005 Report | Commission for Africa*. www.commissionforafrica.info/2005-report

Conteh, F. M., & Maconachie, R. (2019). Spaces for contestation: The politics of community development agreements in Sierra Leone. *Resources Policy, 61*, 231–240.

Crane, A., & Matten, D. (2010). *Business Ethics: Managing Corporate Citizenship and Sustainability in the Age of Globalization*. United Kingdom.

Crane, A., McWilliams, A., Matten, D., Moon, J., & Siegel, D. S. (2008). *The Oxford handbook of corporate social responsibility*. Oxford University Press.

Crowther, D., & Aras, G. (2008). Governance and sustainability: An investigation into the relationship between corporate governance and corporate sustainability. *Management Decision, 46*(3), 433–448.
Crowther, D. Seifi, S. & Wond, T. 2019. Responsibility and governance: The twin pillars of sustainability. In: Crowther, D. Seifi, S. & Wond, T, Responsibility and governance, pp. 1–13.
Dahlsrud, A. (2008). How corporate social responsibility is defined: An analysis of 37 definitions. *Corporate Social Responsibility and Environmental Management, 15*(1), 1–13.
Datzberger, S. (2014). *Peacebuilding and the depoliticisation of civil society: Sierra Leone [2002–2013]. Doctoral Thesis*. London School of Economics and Political Science.
Deegan, C. (2002). The legitimising effect of social and environmental disclosures—A theoretical foundation. *Accounting, Auditing and Accountability Journal, 15*(3), 282–311.
Desjardin, R. (2000). *Business, ethics and the environment*. Pearson.
Dicken, P. (1998). *Global shift: Transforming the world*. Sage.
DiMaggio, P. J., & Powell, W. W. (1983). The iron cage revisited: Institutional isomorphism and collective rationality in organizational fields. *American Sociological Review, 48*, 147–160.
Donaldson, T., & Dunfee, T. W. (1994). *Toward a unified conception of business ethics*. Oxford University Press.
Dowling, J. B., & Pfeffer, J. (1975). Organisational legitimacy: Social values and organizational behaviour. *Pacific Sociological Review, 18*(1), 122–136.
Elkington, J. (1997). *Cannibals with forks: The triple bottom line of 21st century business*. Earthscan.
Emeseh, E. (2009). Social responsibility in practice in the oil producing Niger Delta: Assessing corporations and Government's actions. *Journal of Sustainable Development in Africa, 11*(2), 113–125.
Emeseh, E., & Songi, O. (2014). CSR, human rights abuses and sustainability report accountability. *International Journal of Law and Management, 56*(2), 136–151.
Eweje, G. (2007). Multinational oil companies' CSR initiatives in Nigeria. *Managerial Law, 49*(5/6), 218–235.
Ferguson, J., Sales de Aguiar, T. R., & Fearfull, A. (2016). Corporate response to climate change: Language, power and symbolic construction. *Accounting, Auditing & Accountability Journal, 29*(2), 278–304.
Filho, L., & Brandli, L. (Eds.). (2016). *Engaging stakeholders in education for sustainable development at university level*. Springer.
Fieldhouse, D. K. (1986). *Black Africa 1945–1980: Economic decolonization and arrested development*. Routledge.
Fombrun, C. J., & Van Riel, C. B. M. (1997). The reputational landscape. *Corporate Reputation Review, 1*(1–2), 5–13.
Frederick, W. C. (1986). Toward CSR3: Why ethical analysis is indispensable and unavoidable in corporate affairs. *California Management Review, 28*(2), 126–141.
Frederick, W. C. (1994). From CSR1 to CSR2: The maturing of business and society thought. *Business and Society, 33*(2), 150–164.
Freeman, R. E. (1984). *Strategic management: A stakeholder approach*. Pitman.
Friedman, M. (1962). *Capitalism and freedom*. University of Chicago Press.
Friedman, M. (1970, September 13). The social responsibility of business is to increase its profit. *The New York Times Magazine*.
Frynas, J. G. (2005). The false promise of corporate social responsibility: Evidence form multinational oil companies. *International Affairs, 81*(3), 581–598.
Frynas, G. (2009). *Beyond corporate social responsibility*. Oxford University Press.
Garriga, E., & Mele, B. (2004). Corporate social responsibility theories: Mapping the territory. *Journal of Business Ethics, 53*(1–2), 51–71.
Gokulsing, R. D. (2011). CSR matters in the development of Mauritius. *Social Responsibility Journal, 7*(2), 218–233.

Gond, J.-P., & Moon, J. (2011). Corporate social responsibility in retrospect and Prospect: Exploring the life-cycle of an essentially contested concept. In *No. 59–2011 ICCSR research paper series. Research paper series International Centre for Corporate Social Responsibility*. University of Nottingham.

Grober, U. (2012). *Sustainability: A cultural history*. Green Books.

Guthrie, J., & Parker, L. D. (1989). Corporate social reporting: A rebuttal of legitimacy theory. *Accounting and Business Research, 19*, 343–352.

Hayward, S., Fowler, E., & Steadman, L. (2000). *Environmental quality 2000: Assessing Michigan and America at the 30th anniversary of earth day*. Mackinac Center for Public Policy and the Pacific Research Institute for Public Policy.

Hemmati, M. (2002). *Multi-stakeholder processes for governance and sustainability: Beyond deadlock and conflict*. Earthscan Publications.

Honke, J. (2012). Land grabbing by mining companies: Local contentions and state reconfiguration in south Kivu (DRC). In A. Ansoms & T. Hilhorst (Eds.), *Losing your land: Dispossession in the Great Lakes*. Oxford University Press.

Hooghiemstra, R. (2000). Corporate communication and impression management—New perspectives why companies engage in corporate social reporting. *Journal of Business Ethics, 27*(1–2), 55–68.

Idemudia, U. (2014). Corporate social responsibility and development in Africa: Issues and possibilities. *Geography Compass, 8*(7), 421–435.

Idemudia, U. (2010). Corporate social responsibility and the rentier Nigerian state: Rethinking the role of government and the possibility of corporate social development in The Niger Delta. *Canadian Journal of Development Studies, 30*(1–2), 131–153.

Idemudia, U. (2011). Corporate social responsibility and developing countries: Moving the critical CSR research agenda in Africa forward. *Progress in Development Studies, 11*(1), 1–18.

Idemudia, I., & Ite, U. E. (2006). Corporate-community relations in Nigeria's oil industry: Challenges and imperatives. *Corporate Social Responsibility and Environmental Journal, 13*(4), 194–206.

Ihlen, O., & Roper, J. (2011). Corporate reports on sustainability and sustainable development: 'We have arrived'. *Sustainable Development Sustainable Development, 4*(6), 1–10.

Ioannou, I., & Serafeim, G. (2014). *The consequences of mandatory corporate sustainability reporting: Evidence from four countries*. Working Paper 11-100. Harvard University.

Ite, U. E. (2004). Multinationals and corporate social responsibility in developing countries: A case study of Nigeria. *Corporate Social Responsibility and Environmental Journal, 11*(1), 1–11.

Jones, T. M. (1995). Instrumental stakeholder theory: A synthesis of ethics and economics. *Academy of Management Review, 20*(2), 404–437.

Kamara, H., & Nwagbara, U. (2015). Corporate social responsibility (CSR) leadership and poverty reduction: The case of Nigeria and Sierra Leone. *Economic Insights–Trends & Challenges, IV*(LXVII) 2/2, 21–30.

Kang, N., & Moon, J. (2012). Institutional complementarity between corporate governance and corporate social responsibility: A comparative institutional analysis of three capitalisms. *Socio-Economic Review, 10*, 85–108.

Karl, T. L. (1997). *The paradox of plenty: Oil booms and petro-states*. University of California Press.

Katsos, J. E., & Forrer, J. (2014). Business practices and peace in post-conflict zones: Lessons from Cyprus. *Business Ethics., 23*(2), 154–168.

Kleine, A., & Hauff, M. (2009). Sustainability-driven implementation of corporate social responsibility: Application of the integrative sustainability triangle. *Journal of Business Ethics, 85*(3), 517–569.

Kolk, A., & Lenfant, F. (2013). Multinationals, CSR and partnerships in central African conflict countries. *Corporate Social Responsibility and Environmental Management, 20*(1), 43–54.

Konteh, W. (1997). *Forest resource management in Sierra Leone: A critique of policy formulation and implementation*. Unpublished PhD Thesis. University of Leeds.

KPMG. (2017). *KPMG survey of corporate social responsibility reporting*. Accessed June, 2023, from https://kpmg.com/xx/en/home/insights.html

Kreps, T. (1940). *Measurement of the social performance of business*. United States Government.

Lantos, G. P. (2001). The boundaries of strategic corporate social responsibility. *Journal of Consumer Marketing, 18*(7), 595–632.

Lockett, A., Moon, J., & Visser, W. (2006). Corporate social responsibility in management research: Focus, nature, salience and sources of influence. *Journal of Management Studies, 43*(1), 115–136.

Loew, T., Ankele, K., Braun, S., & Clausen, J. (2004). *Significance of the CSR debate for sustainability and the requirements for companies*. Federal Ministry for the Environment, Nature Conservation and Nuclear Safety.

Lumley, S., & Armstrong, P. (2004). Some of the nineteenth century origins of the sustainability concept. *Environment, Development and Sustainability, 6*(3), 367–378.

Maconachie, R. (2009). Diamonds, governance and 'local' development in post-conflict Sierra Leone: Lessons for artisanal and small-scale mining in sub-Saharan Africa? *Resources Policy, 34*(1), 71–79.

Maconachie, R. (2008). New agricultural frontiers in post-conflict Sierra Leone? Exploring institutional challenges for wetland management in the Eastern Province. *The Journal of Modern African Studies, 46, 2*, 235–266.

Maconachie, R. (2016). The extractive industries, mineral sector reform and post-conflict reconstruction in developing countries. *Extractive Industries and Society, 3*(2), 313–315.

Maconachie, R. 2012. The diamond area community development fund: Micropolitics and community-led development in post-war Sierra Leone. P. Lujala & S.A. Rustad. : Routledge. High-Value Natural Resources and Peacebuilding, pp. 261–273.

Maconachie, R., Binns, T., Tengbe, P. R., & Johnson. (2006). Temporary labour migration and sustainable post-conflict return in Sierra Leone. *GeoJournal, 67*, 223–240.

Margolis, J. D., & Walsh, J. P. (2003). Misery loves companies: Rethinking social initiatives by business. *Administrative Science Quarterly, 48*, 268–305.

Matten, D., & Crane, A. (2005). Can corporations be citizens? Corporate citizenship as a metaphor for business participation in society. *Business Ethics Quarterly, 15*(3), 429–453.

Mele, V., & Schepers, D. (2013). E pluribus unum? Legitimacy issues and multi-stakeholder codes of conduct. *Journal of Business Ethics, 118*(3), 561–576.

McGuire, J. W. (1963). *Business and society*. MaGraw-Hill.

Mebratu, D. (1998). Sustainability and sustainable development: Historical and conceptual review. *Environmental Impact Assessment Review, 18*(6), 493–520.

Miles, S. (2012). Stakeholder: Essentially contested or just confused? *Journal of Business Ethics, 108*, 285–298.

Mitnick, B. M. (2000). Commitment, revelation, and the testimonies of belief: The metrics of measurement of corporate social performance. *Business and Society., 39*, 419–465.

Newell, P. (2005). Citizenship, accountability and community: The limits of the CSR agenda. *International Affairs, 81*(2), 541–557.

Noyoo, N. (2010). Linking corporate social responsibility and social policy in Zambia. In P. Utting & J. C. Marques (Eds.), *Corporate social responsibility and regulatory governance: Towards inclusive development?* (pp. 105–123). Palgrave Macmillan.

Obi, C. (2010). Oil as the 'curse' of conflict in Africa: Peering through the smoke and mirrors. *Review of African Political Economy, 37*(123), 483–495.

Okoye, A. (2009). Theorising corporate social responsibility as an essentially contested concept: Is a definition necessary? *Journal of Business Ethics, 89*(4), 613–627.

Orlitzky, M. (2008). Corporate social performance and financial performance: A research synthesis. In A. Crane, A. McWilliams, D. Matten, J. Moon, & D. Siegel (Eds.), *The Oxford handbook of corporate social responsibility* (pp. 24–38). Oxford University Press.

Orlitzky, M., Schmidt, F. L., & Rynes, S. L. (2003). Corporate social and financial performance: A meta-analysis. *Organisation Studies, 24*(3), 403–441.

Pan, X., Chen, X., & Ning, L. (2018). The roles of macro and micro institutions in corporate social responsibility (CSR): Evidence from listed firms in China. *Management Decision, 56*(5), 955–971.

Parris, T. M., & Kates, R. W. (2003). Characterising and measuring sustainable development. *Annual Review of Environmental Resources, 28*(13), 113–128.

Parson, T. (1960). *Structure and process in modern society*. Free Press.

Porto, N. (2010). Sierra Leone. In W. Visser & N. Tolhurst (Eds.), *The world guide to CSR: A country-by-country analysis of corporate sustainability and responsibility* (pp. 349–359). Greenleaf Publishing.

Posnikoff, J. F. (1997). Disinvestment from South Africa: They did well by doing good. *Contemporary Economic Policy, 15*(1), 76–86.

Purvis, B., Mao, Y., & Robinson, D. (2018). Three pillars of sustainability: In search of conceptual origins. *Sustainability Science, 14*, 681–695.

Ragodoo, N. (2009). CSR as a tool to fight against poverty: The case of Mauritius. *Social Responsibility Journal, 5*(1), 19–33.

Raman, R., & Lipschutz, R. (2010). *Corporate social responsibility: Comparative critiques*. Palgrave.

Renouard, C. (2011). Corporate social responsibility, utilitarianism, and the capabilities approach. *Journal of Business Ethics, 98*(1), 85–97.

Richey, L. A., & Ponte, S. (2014). New actors and alliances in development. *Third World Quarterly, 35*(1), 1–21.

Samy, M., Odemilin, G., & Bampton, R. (2010). Corporate social responsibility: A strategy for sustainable business success. An analysis of 20 selected British companies. *Corporate Governance: The International Journal of Business in Society, 10, 2*, 203–217.

Scherer, A. G., & Palazzo, G. (2011). The new political role of business in a globalised world: A review of a new perspective on CSR and its implications for the firm, governance, and democracy. *Journal of Management Studies, 48*(4), 899–931.

Secchi, D. (2007). Utilitarian, managerial and relational theories of corporate social responsibility. *International Journal of Management Reviews, 9*(4), 347–373.

Schaltegger, S., & Wagner, M. (2006). *Managing the business case of sustainability. Sustainability performance, competitiveness and business success: Frameworks, empirical results and management approaches*. Greenleaf.

Sethi, S. P. (2007). Investing in socially responsible companies is a must for public pension funds—Because there is no better alternative. *Journal of Business Ethics, 56*, 99–129.

Silberhorn, D., & Warren, R. C. (2007). Defining corporate social responsibility: A view from big companies in Germany and the UK. *European Journal Business Review, 19*(5), 352–372.

Siltaoja, M. E. (2009). On the discursive construction of a socially responsible organisation. *Scandinavian Journal of Management, 25*, 191–202.

Siltaoja, M. E., & Onkila, T. J. (2013). Business in society or business and society: The construction of business–society relations in responsibility reports from a critical discursive perspective. *Business Ethics: A European Review, 22*(4), 357–373.

Smillie, I., Gberie, L., & Hazleton, R. (2000). *The heart of the matter: Sierra Leone, diamonds & human security (complete report)*. Partnership Africa Canada.

Suchman, M. C. (1995). Managing legitimacy: Strategic and institutional approaches. *Academy of Management Review, 20*(3), 571–610.

Thompson, J. D. (1967). *Organisations in action: Social science bases of administrative theory*. McGraw-Hill.

Tonelson, A. (2002). *The race to the bottom : Why a worldwide worker surplus and uncontrolled free*. Westview Press.

Ullman, A. A. (1985). Data in search of a theory: A critical examination of the relationships among social performance, social disclosure, and economic performance of U.S. firms. *Academy of Management Review, 10, 3*, 540–577.

UNCTAD. (2000). *World investment report 2000 cross-border mergers and acquisitions and development*. United Nations.
United Nations. (2006). *The millennium development goals report*. United Nations.
Visser, W. (2013). *The age of responsibility: CSR 2.0 and the new DNA of business*. Wiley.
Visser, W. (2006). Revisiting Carroll's CSR pyramid an African perspective. In E. R. Pedersen & M. Huniche (Eds.), *Corporate citizenship in developing countries* (pp. 29–56). Copenhagen Business School Press.
Visser, W., McIntosh, M., & Middleton, C. (2006). *Corporate citizenship in Africa: Lessons from the past; paths to the future*. Greenleaf Publishing.
Waddock, S. A., & Graves, S. B. (1997). The corporate social performance-financial performance link. *Strategic Management Journal, 18*(4), 303–319.
Whelan, G. (2012). The political perspective of corporate social responsibility: A critical research agenda. *Business Ethics Quarterly, 22*(4), 709–737.
Williams, S., & Preston, D. (2018). Working with values: An alternative approach to win-win. *International Journal of Corporate Strategy and Social Responsibility, 1*(4), 302–319.
Wilson, S. A. (2015). Corporate social responsibility and power relations: Impediments to community development in post-war Sierra Leone diamond and rutile mining areas. *The Extractive Industries and Society, 2*(4), 704–713.
Windsor, D. (2006). Corporate social responsibility: Three key approaches. *Journal of Management Studies, 43*(1), 0022–2380.
Wood, D. J. (2010). Measuring corporate social performance: A review. *International Journal of Management Review, 12*(1), 50–84.
World Commission on Environment and Development (WCED). (1987). *Our Common Future*. Oxford University Press.
Wright, P., & Ferris, S. (1997). Agency conflict and corporate strategy: The effect of divestment on corporate value. *Strategic Management Journal, 18*, 77–83.
Zerk, J. A. (2006). *Multinationals and corporate social responsibility: Limitations and opportunities in international law*. Cambridge University Press.
Zwetsloot, G. I. J. M. (2003). From management systems to corporate social responsibility. *Journal of Business Ethics, 44*, 201–208.

Henry Yamba Kamara, PhD is an agribusiness sustainable development and CSR expert based in Sierra Leone. He is holding the PhD in Sustainable Development with a concentration in CSR from Cardiff Metropolitan University; an Executive Education in Trade Policy from Harvard University; MSc from Liverpool John Moores University in Corporate Governance, and an MSc in the Management of Quality Excellence from the University of Leicester. Dr. Kamara is a Certified Lead Assessor in ISO 9001, a Certified Lead Auditor in Roundtable Sustainable Palm Oil (RSPO), a Certified Master trainer in Cashew Production, and can assist organization to achieve independent external ISO9001 Quality Management Systems and RSPO Certification. He is a published consultant and worked as Managing Director of the Sierra Leone Produce Marketing Company, In-Country Facilitator of Tropical Forest Alliance (TFA2020) Sierra Leone, Commissioner National Commission for Privatization (NCP) Sierra Leone, Director of Export Development Sierra Leone Investment & Export Promotion Agency, Head of Quality Management Sierra Leone Standards Bureau and National Value Chain, Sectors Specialist, United Nations Industrial Development Organization (UNIDO), Team Leader, KPMG International Development Advisory Services (IDAS) for the Sierra Leone Agribusiness Development Fund (SLADF), and Grants Management Specialist for the Smallholder Commercialization and Agribusiness Development Project (SCADeP). He has an extensive knowledge in Project Management, Research, Resource Mobilization, Grants Management, and Environmental and Social Safeguards and has participated as an independent consultant in the implementation of national and regional projects supported by UNIDO, ITC, UNCTAD, AfDB, The World Bank, IFC, KPMG, Solidaridad WA, Proforest UK, EU, and ECOWAS.

Uzoechi Nwagbara, PhD is a professor of management at ICT University. He has BA and MA in English as well as MSc and PhD in Human Resource Management and Management, respectively, from the University of Wales, and MEd in Higher Education as well as PG Cert in Academic Practice from the University of Sunderland, UK. He is a Fellow of the Higher Education Academy (FHEA); Member Chartered Management Institute (MCMI); Member Institute of Leadership & Management (MInstLM); and Member Community Development Exchange (MCDE); and also affiliated with British Academy of Management (BAM), European Association for the Study of Literature, Culture, and the Environment (EASLCE), Postcolonial Studies Association (PSA), and Royal African Society (RAS). Dr Nwagbara has over 35 years of teaching experience and has taught in higher education for upwards of 25 years. He is external examiner for University of Cumbria and University of Bolton. He has supervised over a dozen doctoral students to completion and published well over 100 publications in high impact journals in management, leadership, CSR, sustainability, employment relations, and human resource management. Dr Nwagbara has examined PhD students internally and externally in various universities including University of Bedfordshire, Cardiff Metropolitan University, and University of the West of Scotland amongst others. He holds Visiting Teaching Positions at the University of Sunderland in London, Coventry University London, London School of Commerce, University of the West of Scotland, The University of Law UK, University of Worcester, University of Wales Trinity Saint David, and University of Plymouth as well as Visiting Professorship (management), Coal City University. He is on the editorial boards of many peer-reviewed, international journals including *Economic Insights—Trends & Challenges*, *Journal of Pan African Studies* and *Journal of Sustainable Development in Africa*.

Chigozie A. Ugwoji, BSc, MA, MSc is a doctoral candidate at Brunel University, London, UK. She is currently a lecturer at University of Sunderland in London and is a certified social worker with many years of experience. Chigozie's areas of specialties include leadership, sustainable development, social work, management, and corporate governance.

Ngozi Ibeawuchi, BA, MA is a doctoral candidate at the University of Greenwich. She is currently lecturing at Coventry University London and Regent College, UK. Her areas of expertise include leadership, special education, and CSR and she has published in several journals. Her doctoral focus is on special education and the role played by transformational leadership in the process of making educational leadership more inclusive.

Chapter 10
An Analysis of Corporate Social Responsibility (CSR)-Related Objectives Enshrined in the "National Oil and Gas Policy for Uganda, 2008": A Policy Perspective Paper

David Katamba, Bosco Amerit, Maureen Basuuta, Swithern B. Tumwine, and Ngoma Muhammed

Background

Corporate Social Responsibility (CSR) in the Era of Natural Resources Management

Corporate Social Responsibility (CSR) is the continuing voluntary commitment by an organization to demonstrate its responsibility and accountability for its actions toward different stakeholders as it pursues its economic interests (Katamba, 2018). Very often, the CSR activities claimed to have been done by companies or the impact of CSR activities, are verified through a "*Social Compliance Audit.*" The activities should be voluntarily conducted in a way that reflects a continuing process in which organizations (e.g., O&G sector players) endeavor to protect the health, safety, and rights of their employees, the community and the environment in which they operate, and the lives and communities of workers in their supply and distribution chains.

From this understanding, it becomes clear that it is everyone's (businesses inclusive) turn to take a proactive responsibility (Business call to action, 2013). For example, borrowing from the Extractive Industries Transparency Initiative (EITI)

D. Katamba (✉) · B. Amerit · M. Basuuta · S. B. Tumwine
Makerere University Business School (MUBS), Kampala, Uganda
e-mail: dkatamba@mubs.ac.ug; dkatamba@uccsri.com; bamerit@mubs.ac.ug; bamerit@mubs.ac.ug

N. Muhammed
Living Earth Uganda, Kampala, Uganda
e-mail: mngoma@mubs.ac.ug

U. Nwagbara et al. (eds.), *Corporate Social Responsibility Disclosure in Developing and Emerging Economies*, CSR, Sustainability, Ethics & Governance, https://doi.org/10.1007/978-3-031-61976-2_10

(see, EITI Standard of 2019), we see attempts to place the debate and policy direction on CSR in the context of natural resources income /wealth redistribution as a contributor to the "*Future we want*" globally (see, Brundtland Commission, 1987). For example, Long (2022), in his attempts to investigate "*CSR and reconciliation with Indigenous peoples in Canada,*" established that the CSR discourse needs to accommodate the "free prior and informed consent" of Indigenous peoples and their spiritual values and knowledge *vis-à-vis* the land for resource extractive practices to edge toward being socially responsible when they occur on Canadian ancestral territories. However, in the context of Oil and Gas (O&G) in Uganda (a Low-Income Country—LIC), this is not freely happening, leading to emerging potential signs of the "*Oil curse*" in the Albertine (Uganda's geological zone with concentration of O&G discovery, exploration, and production). For example, much as the exploration companies, notably TotalEnergies E&P, have tried their best to ensure community harmony, there are disgruntlement in Bunyoro (part of Albertine region) of how to share returns, biodiversity loss, food insecurity, endangering gender, and decline in non-O&G local economic activities (Rwengabo, 2017; Kuteesa, 2014). From the CSR perspective, we postulate that if this problem is not fully averted earlier, will suffocate attainment of SDGs (12, 7, 8, 5, 1) in Uganda since "Social Compliance/CSR" seems to be undervalued by Uganda's O&G authorities (particularly Ministry of Energy and Mineral Development—MoEMD, and Petroleum Authority of Uganda—PAU). This undervaluing is evident in the way these Authorities are focused on demanding an Environmental Impact Assessment (EIA) from O&G players before issuing them a "Legal License" to operate. This practice was named by Ross (2012) as inadequate to prevent the "*Oil Curse*." Katamba et al. (2014) suggested that one of the solutions to this problem is adding priority on demanding O&G players to detail strategies of how they are to engage in CSR, which gives them a "Social License" in addition to a "Legal License" to operate. However, O&G players are not mandated by the Ugandan laws and policies to show how they engage in CSR practices that promote Social Compliance [– voluntary activities to guard economies against the signs of Oil Curse). Worst still, no data exists in Uganda's O&G sector about CSR activities in the context of SDGs. Previous related studies (Rwengabo, 2017; Kuteesa, 2014) focused on energy efficiency, and economic viability of O&G quantities but not CSR for social compliance thus creating a knowledge gap warranting this publication to inform policy shifts in Uganda.

Petroleum Data in Uganda

The discovery of Oil and Gas (O&G) in Uganda is traced back to the early 1960s around Lake Albert. However, the actual commercial engagement in O&G related activities could not start until recently in the late 2000s due to lack of petroleum data. Petroleum data refers to all the data relied upon by Government, oil companies or other actors/stakeholders to determine the existing petroleum potential, extent of reserves, and amount recoverable. It comprises geological data, gravity and magnetic data, seismic data, and well data. Since the 1980s, Government and private companies have been collecting petroleum data. Government collected preliminary geological and geophysical data to identify the petroleum potential and market-demarcated blocks. Following this, private oil companies came to conduct further reconnaissance studies and later oil exploration. The discovery of oil in commercial quantities in Uganda in 2006 has thus triggered more activities in the upstream and downstream activities (Mwakali & Byaruhanga, 2018).

Hence, following the availability of working petroleum data, Oil and Gas (O&G) Policies and Regulations have been enacted in Uganda to guide further engagement into activities (downstream and upstream) related to this subsector. That is, to majorly guide the exploration, production, and issuance of licenses related to any O&G related business and activities. The major players so far issued with "production" and/or "Exploration" licenses are International Oil Companies (IOCs). These include Total Exploration and Production (TotalEnergies E&P), China National Offshore Oil Corporation (CNOOC), and Tullow Plc. These oil companies have been involved in exploration and development activities in the Albertine Graben. As a result of the exploration and development activities in the Albertine Graben, the Minister of Energy and Mineral Development (MEMD) granted eight (8) Petroleum Production Licenses for oil fields in Exploration Area (EA2) and Exploration Area 1 (EAI) to Tullow, Total and CNOOC as joint venture partners on August 30th, 2016, marking the start of the production phase; the first oil had been expected in the year 2020, though due to unforeseen constraints, this has not been realized yet.

Rationale

Recently (March 24th, 2023), Uganda's Ministry of Finance, Planning, and Economic Development (MoFPED) released and launched Uganda's Public Investment Financing Strategy (PIFS).[1] The PIFS, which was launched at a time when development is being defunded, explores additional financing from domestic, external, emerging, and innovative financing options to meet the increasing development aspirations embedded in the National Development Plans (NDP) and the Sustainable Development Goals. It also provides a framework for collaboration with the private sector in the implementation and financing of public investments. In this respect, the PIFS places significant consideration on embracing Corporate Social Responsibility (CSR) as one of the turning points that can raise resources and integrate everyone in the path to national development. Specifically, under Sect. 2.2.13, government points that CSR in Uganda is not formalized or institutionalized

[1] The launch of the PIFS comes at a time when the world is racing towards the 2030 deadline for achieving SDGs, but also when the global economy is grappling with the aftereffects of COVID-19, the war in Ukraine, rising geopolitical tensions and a worsening climate change crisis. https://www.finance.go.ug/sites/default/files/Publications/PUBLIC%20INVESTMENT%20FINANCE%20STRATEGY%20FINAL.pdf

and the government is yet to develop a CSR policy framework. CSR activities remain largely ad hoc and unpredictable. Thus, as a mitigation measure, the PIFS clearly mentions that:

> … Government will enhance philanthropy activities through the following strategies: … Create a national legal framework for Corporate Social Responsibility (CSR) that clearly defines the roles and expectations for businesses, government institutions, development partners and other stakeholders…, (PIFS, pg. 27).

Henceforth, this policy perspective paper digs into Uganda's main O&G policy (*National Oil and Gas Policy for Uganda,* 2008) which can be reviewed, to strengthen PIFS' ambition on CSR and national development. The policy paper's emphasis has been placed on O&G players' CSR/social compliance policy environment since these players can be strategically good partners in Uganda's socio-economic development. Therefore, in addition to the four (4) O&G Exploration and Production companies (upstream players—CNOOC, TotalEnergies E&P, UNOC), there are also 19 O&G Downstream players as of January 1st, 2021, see Table 10.1. These offer services related to O&G waste management, distribution, and marketing & sale of petroleum products. Hence, this policy document's concern is that the issuance of these licenses is largely driven by production and presentation of an "Environmental Impact Assessment (EIA)" report. We would henceforth want to see a big integration of Corporate Social Responsibilities (CSR) in the form of Social Compliance. That is, from the CSR perspective of upstream O&G companies and their allies in downstream activities, this "Policy Perspective Paper" is intended to assess the impact of the current CSR entry points indicated in "*National Oil and Gas Policy for Uganda, 2008*" in order to provide room for scaling up O&G players CSR interventions and activities that engage women, youth, and the elderly persons in efforts to promote social-economic inclusion, environmental responsibility, and community development. The Government of Uganda (GoU) will use this Policy Analysis to devise policy shifts. Notably, shift from overdependence on Environmental Impact Assessment (EIA) reports as the major prerequisite to be

Table 10.1 Downstream O&G players in Uganda

O&G Downstream players as of January 1st, 2021
A total of nineteen (19) downstream players were active at the time of compiling this policy perspective paper. These are stratified as follows:
• Two (2) Environmental and Waste Management Services companies: Epsilon; Air, Water, Earth (AWE).
• Four (4) Camp Management and Catering companies: Strategic Logistics (SLL), MSL Logistics, Gulf Catering Company Services, and Equator Catering Ltd.
• Five (5) Transport and Logistics companies: Threeways Shipping, Spedag-Interfreight, SDV Transami, Quantum Associates Oil & Gas Logistics, and DHL.
• Two (2) Inspection Services companies: Inspecta International and Lloyds British Testing Uganda.
• Six (6) Other (Insurance, Distribution, Associations, Training): Uganda Chamber of Mines and Petroleum; Petroleum Skills Uganda Ltd., Habib Oil, General Machinery, Association of Uganda Oil and Gas Service providers (AUOGS), and AON Risk.

Source: Petroleum Authority of Uganda (PAU) website https://www.pau.go.ug/

granted (or renew license) to operate in the Albertine and/or in Uganda's O&G sub-sector, to also demanding comprehensive and operationally relevant integrated CSR plans that enhance social compliance. Additionally, this Policy Analysis is to guide policy makers to improve their capacity in assessing the quality of annual reporting of O&G players by additionally demanding for performance data on CSR and Social Compliance aspects. The Local Government officials/policy implementers in the Albertine, and Ministry of Gender, Labor and Social Development will also use this research output to: (i) monitor the relevance of O&G companies and their allies to policy instruments that promote social inclusion and sustainable development. (ii) Make informed decision about the readiness of these players before entering a Memorandum of Understanding (MoU) for potential CSRR and PPP projects. Lastly, (iii) make strategic plans for social development and gender mainstreaming.

Highlights About the National Oil and Gas Policy for Uganda, 2008

Uganda has a series of policies that govern the Oil and Gas (O&G) subsector, of the energy sector. The main one is "*The National Oil and Gas Policy for Uganda (*2008*)*." This main policy has been preferred for this policy perspective paper because it addresses the entire spectrum of exploration, development, and production of the country's O&G resources more comprehensively than the *Energy Policy of 2002.* Moreover, this policy [*National Oil and Gas Policy*] recognizes Corporate Social Responsibility (CSR) as an important issue that can be embraced by the O&G sector players as they pursue their O&G activities (see Table 10.2). However,

Table 10.2 A quick snapshot of the National Oil and Gas Policy

Policy, legal and regulatory framework of the National oil and gas policy
The National oil and gas policy for Uganda was approved in 2008 to provide general oversight and guidance for the Country's emerging oil and gas sector. The policy highlights the key issues, guiding principles, objectives together with strategies and actions, and institutional responsibilities in developing the sector. In line with the policy, two laws were enacted in 2013 to update the legal framework for the sector. These are: (a) The Petroleum (Exploration, Development and Production) Act 2013 (b) The Petroleum (Refinery, Conversion, Transmission and Midstream Storage) Act 2013 Seven sets of regulations on Technical, Health, Safety, and Environment, National Content and Metering aspects are in place. A Local Content Policy for the Oil and Gas Sector in Uganda was approved by Cabinet in June 2018; 2012 Oil and Gas Revenue Management Policy; The Public Finance Act 2015 was enacted in March 2015 and provides for among others the management of revenues accruing from petroleum activities. These pieces of legislation together with other relevant laws and statutes like those on Environment, Wildlife, Water, Land, and Income Tax, are used in regulating the petroleum sector.

Source: Compiled from "The Oil and Gas Sector in Uganda,"—Petroleum Authority of Uganda https://www.pau.go.ug/laws/

the policy does not define the strategies and actions required to achieve the desirable outcomes in the other sectors because they are best articulated in the respective sector policies. Additionally, the policy promotes positive impacts of O&G activities on other sectors (like agriculture, transport, and education) and simultaneously seeks to mitigate any negative impacts. Our analysis and review of this policy, highlights that under this Policy, there are two objectives that provide guiding principles relating to Social Compliance [*broadly put,* Corporate Social Responsibility—CSR). We thus present these CSR related objectives below, and resultantly a detail of their enabling impact on selected "CSR/Social Compliance" aspects/facets in the proceeding sections (see Table 10.3).

Objectives Related to CSR/Social Compliance and Strategies to Achieve them

From the review of *The National Oil and Gas Policy for Uganda (*2008*),* two objectives were identified as promoters of Corporate Social Responsibility (CSR) /Social Compliance. These are Objective 9 and Objective 10 as presented below with their respective strategies to realize them.

- **Objective 9:** To ensure that oil and gas activities are undertaken in a manner that conserves the environment and biodiversity. Under this are strategies that include to:

  (a) "… Ensure availability of the necessary institutional and regulatory framework to address environment and biodiversity issues relevant to oil and gas activities.
  (b) Ensure presence of the necessary capacity and facilities to monitor the impact of oil and gas activities on the environment and biodiversity.
  (c) Require oil companies and their contractors/subcontractors to use self-regulation and best practices in ensuring environmental protection and biodiversity conservation.
  (d) Require oil companies and any other operators to make the necessary efforts to return all sites on which oil and gas activities are undertaken to their original condition as an environmental obligation…."

  Source: The National Oil and Gas Policy for Uganda (2008).

  The actions to achieving these strategies are stipulated as follows:

(i) "…Upgrading the relevant Environment and Biodiversity legislation to address oil and gas activities.
(ii) Strengthening institutions with a mandate to manage the impact of oil and gas activities on the environment and biodiversity.

Table 10.3 Differences between EIA, ESIA, and CSR/Social Compliance

	Environment Impact Assessment (EIA)	*Environmental and Social Impact Assessment (ESIA)*	*CSR /Social compliance*
Meaning	EIA is a systematic examination conducted to determine whether, or not, a project will have any adverse impacts on the environment	It is a thorough study of the impact of the investment project on the environment and the social fabric of the local communities	A continuous process and activities in which organizations engage to protect the health, safety and rights of their employees, the community and environment in which they operate, the lives and communities of workers in their supply and distribution chains
Objective	• To identify, *predict and evaluate* the economic, environmental, and social impact of development activities • To provide information on the environmental consequences of decision-making • To promote environmentally sound and sustainable development through the identification of	• To predict environmental and social consequences that a *future project*/intervention might entail • To understand who is likely to be impacted and how • To understand the affected communities • To identify and assess potential social impacts • To develop management measures to mitigate adverse impacts and enhance benefits	• To *put in action* (operationalize) what the ESIA promises or safeguards that were promised in the EIA • To audit the CSR activities, undertake and provide an opinion regards the sustainable development impact created by such activities
When is it done?	• Before project implementation to show the potential *negative environmental impacts* of any project / investment	• Before project implementation and propose measures to mitigate *potential negative environment and social impacts* **NB:** The ESIA is a higher version of EIA. This is evident from the review of EACOP by the Netherland Commission for Environmental Assessment (NCEA) on behalf of NEMA	Continuously across the lifetime of project implementation /investment

Source: Authors' compilation
Advisory Review of the resubmitted Environmental and Social Impact Assessment for the East Africa Crude Oil Pipeline (EACOP), https://www.eia.nl/docs/os/i72/i7228/7228_website_versie_advisory_report_eacop_uganda_22_october.pdf

(iii) Developing physical master plans, environmental sensitivity maps and oil spill contingency plans for the oil and gas producing region and any transport corridors...."

Source: The National Oil and Gas Policy for Uganda (2008).

Objective 10: To ensure mutually beneficial relationships between all stakeholders in the development of a desirable oil and gas sector for the country. The achievement of this objective is through the achievement of the following strategies:

(a) "…Recognize and promote the different roles of the state, oil companies and other stakeholders.
(b) Ensure that the PSA's entered into have clear obligations about work programs and fiscal obligations.
(c) Provide clear procedures for any framework adjustments, arbitration, and exit points.
(d) Require good and established self-regulation and monitoring mechanism by oil companies.
(e) Ensure that oil companies and their contractors/subcontractors implement the necessary guidelines regarding health and safety at work including provision of protective equipment where necessary.
(f) Encourage the implementation of corporate social responsibility by oil companies.
(g) Provide for availability of information that may be required by stakeholders…."

Source: The National Oil and Gas Policy for Uganda (2008).
To be achieved through the implementation of the following actions:

(i) "…Monitor and evaluate the performance of institutions participating in the development of the oil and gas sector in the country.
(ii) Periodically review the regulatory framework with a view of making any necessary changes.
(iii) Put in place an efficient communication strategy for the oil and gas sector.
(iv) Carry out adequate consultations with stakeholders especially in the oil and gas producing areas…."

Source: The National Oil and Gas Policy for Uganda (2008).

The Impact of National Oil and Gas Policy for Uganda, 2008 from Different CSR Perspectives

Borrowing from the Extractive Industries Transparency Initiative (EITI)[2] to which Uganda is a signatory, impactful Corporate Social Responsibility (CSR) activities are clustered around many facets in the context of Oil and Gas (O&G). It is from these facets that Impact can be seen, felt, monitored, and finally measured. Notable

[2] The Extractive Industries Transparency Initiative (EITI) is an international, multistakeholder initiative that promotes transparency and accountability in the oil, gas, and mining sectors through the

of these, and which this policy perspective paper looks at, include: (i) Landownership, (ii) Physical Planning, (iii) Fiscal and Monetary Policy, (iv) Agriculture and Rural Activities, (v) Impact on Fisheries, (vi) Population Distribution, (vii) Energy Availability, (viii) Employment Opportunities and Patterns, (ix) Environment, (xi) Health, (xii) Education and Research, (xiii) Relations with Neighboring Countries, and, (xiv) National Economy. Let us henceforth examine below how the National Oil and Gas Policy for Uganda, 2008, impacts each of these CSR facets.

Impact on Land Ownership and Use

Implementation of oil and gas activities affected the existing land policies, laws, regulations, and practices. Petroleum exploration, development, production, processing, and transportation impact land ownership and use. During the phase of petroleum exploration, land was used for carrying out surveys and temporary accommodation and this use was for a limited period while the other phases of the petroleum value chain require land to be acquired and held on a long-term basis. Article 237 of the Constitution, the Land Act 16 of 1998, and the National Land Use Policy 2004 together with the Land Sector Strategic Plan 2001–2011 define the existing policies and regulations for land ownership and use in the country. The oil companies and/or Government acquire land for the different aspects of the petroleum value chain like provision of transportation corridors, i.e., pipelines, roads together with that for infrastructure such as wellheads, oil and gas processing facilities, and refineries. Agreements were necessary to be entered into with the landowners to cover such aspects as compensation for their land surface interests. Under such an obligation the Government is required, where necessary and in accordance with the Constitution to acquire land in the public interest to support implementation of oil and gas activities.

Impact on Physical Planning

The urban centers resulting from the development of oil and gas activities need to be well planned to meet the challenges of urbanization arising from the oil trade activities. This policy is aligned with the other relevant national policies and seeks to ensure organized urbanization in and around the Albertine Graben *(Whether this has been done, is a question that field work can answer).* To achieve this objective, consideration was to be made to declare the areas where oil and gas activities would be concentrated in special planning areas to pave way for their physical planning.

---

disclosure of government and company data in resource-rich countries. https://eiti.org/documents/eiti-principles

This entails the identification of potential urban and growth centers around the Graben and the proactive planning for these centers with the oil and gas activities in mind. Planning for these areas could be phased in line with the progress of these activities. Support was to be directed to efforts on the design and implementation of physical planning in pace with the development phases of oil and gas activities. Community participation was ensured in both planning and implementation. All infrastructure and other spatial developments were required to be phased in accordance with prepared structure and detailed plans *(Whether this has been done, is a question that field work can answer).*

Impact on Fiscal and Monetary Policy

This policy recognizes the important relationship between fiscal and monetary policy and the country's oil and gas extraction policies and shall promote their harmony. It further recognizes the need to enhance fiscal discipline over any revenues generated from oil and gas activities. The policy supports priorities of reducing the fiscal deficit and prioritizing spending that promotes an increase in productivity and competitiveness of other sectors in the economy. It discourages the use of oil and gas resources to increase public spending in areas that do not add value to the economy.

Impact on Agriculture and Rural Activities

This policy shall, in collaboration with other relevant national policies, support significant and positive impact of oil and gas activities on the country's agricultural sector in line with the objectives of the country's Plan for Modernization of Agriculture (PMA). This could be achieved through the enhancement of specific areas with higher multiplier effects, like provision of energy services and infrastructure for rural agro-processing industries, together with the increased use of modern and mechanized agriculture through the provision of more affordable inputs like fuels and fertilizers. On the other hand, the policy seeks to promote mitigation measures against people abandoning agricultural activities and migrating to the oil and gas sector. Mitigation measures are anticipated to include using oil and gas activities to support agricultural zoning, agro-processing, agro-marketing, agro-inputs, and pro-poor agro-biotechnologies *(Whether this has been done, is a question that field work can answer).*

Impact on Fisheries

Oil and gas exploitation could have had some effects on the aquatic environment, especially in the fisheries of lakes surrounding the areas where oil and gas activities have been undertaken. Hydrocarbon gases, if released within water may rapidly penetrate fish, especially through the gills and disturb their main functional systems. Increasing temperature and reduced levels of oxygen certainly affect fish metabolism resulting in a reduction of their resistance to many organic and inorganic contaminants. On the other hand, oil and gas activities have and continue to lead to improved accessibility to fishing communities resulting in improved marketing opportunities of fish for the fishermen. It is, therefore, recognized that whereas some of the effects of oil and gas activities on fisheries will be positive others may be negative. This policy seeks to promote positive impacts while ensuring that efforts are made to mitigate the negative effects *(Whether this has been done, is a question that field work can answer)*.

Impact on Population Distribution

About 12% of Uganda's population live in urban areas to escape rural poverty and with the belief that cities, municipalities, and towns offer greater opportunities to improve their social and economic livelihoods. Smallholder agriculture mainly in the rural areas supplies the food upon which the urban populations depend. Experience from some countries shows that oil and gas-producing regions may attract labor and threaten other sources of productivity thus leading to the abandonment and collapse of other sectors of the economy. It is therefore possible that large numbers of people may move to the Albertine Graben where oil and gas activities are likely to be concentrated despite the area being largely wildlife and biodiversity protected area. This policy recognizes the need to guide population movements and settlements triggered by oil and gas activities. Local industries supplying the oil and gas activities as much as possible ought to be spread throughout the country to avoid concentration around typically oil and gas centers of activity. This policy sets out to support enforcement of regulations restricting population movements and settlements in the protected wildlife and biodiversity areas. Only a minimum of required infrastructure like access roads and installations together with the required workforce are to be allowed into the protected areas *(Whether this has been done, is a question that field work can answer)*.

Impact on Energy Availability

The main sources of energy in Uganda are: 93% biomass, 6% petroleum products, and 1% electricity. The petroleum products used in the country are gas oil (diesel), petrol, aviation fuel, fuel oil, kerosene, lubricants, and bitumen. All are imported thereby exposing the country's economy to fluctuating and exorbitant prices together with a significant level of uncertainty regarding their availability. To reduce the impact of oil price shocks and the erratic supply of imported petroleum products on the economy, this policy seeks to support the use of the country's oil and gas resources to meet domestic demand, export the surplus within the region and subsequently overseas. The use of the country's oil and gas resources meant to substitute imported products is intended to facilitate value addition to these resources, import substitution of petroleum products, and improved security of supply for petroleum products in the country. Pricing of petroleum products refined from the country's oil and gas resources shall not distort the value of the country's resources through subsidies. Oil and gas resources can be used to support the energy mix. A critical impact on energy availability will be utilized to generate electricity, with a view to addressing the severe electricity shortage in the country resulting from limited hydropower generating capacity, climatic changes, and the increasing demand. This policy supports the use of best lessons and practices in utilizing oil and gas resources to enhance the availability of energy.

Impact on Employment Opportunities and Patterns

Oil and gas activities tend to be capital intensive and mainly generate direct employment for science and engineering professionals and technicians, among others. These activities are not labor intensive and therefore the number of people directly employed on a sustainable basis may not be large. However, the industry provides significant employment opportunities through chain or multiplier effect. Businesses that come into being to provide goods and services to the oil and gas sector will provide employment opportunities. Expansion of other sectors of the economy because of the development of an oil and gas sector will also lead to induced growth of employment opportunities. This policy promotes employment of nationals in the oil industry and in the resulting employment created by the industry's chain or multiplier effect. Transportation and storage of any oil and/or gas produced within the country may lead to the construction of pipelines, roads, rail lines, and storage facilities. This construction will create a large demand for semi-skilled and unskilled labor which should be recruited from the local communities. Significant amounts of cash were also meant to come from compensation claims. This policy promotes the development of appropriate procedures for compensation and provision of employment to local communities *(how effectively this has been done is a matter of field work*). Efforts are to be made to put in place micro-development projects in these

Table 10.4 Case study of how a CSR policy on Oil and Gas exploration can create Jobs

Jobs and oil project
Living Earth Uganda (LEU) with funding from European Union implemented a project in the Albertine in awake of Oil and Gas (O&G) exploration and production activities. The Jobs and Oil project was designed to benefit 11 BTVET institutions in the targeted geographical area, as well as 80 microenterprises /associations /groups. From each of the microenterprises / associations /groups, three main beneficiaries were targeted to be directly imparted with skills (in the areas of financial management, leadership and governance, bookkeeping, functional literacy, vocational skills like soap making, candle making, tailoring, etc.). The aim was to enable them to be "trainers of trainees (ToT)." This would cut costs while ensuring sustainability of the impact created by the Jobs and Oils project. Hence, giving a direct beneficiaries number to be 240 (that is, 3 ToTs X 80) people from enterprises, and an estimated over 2500 youths from BTVET institutions. It should be, however, noted that on average, each of these group / enterprises comprised over 20 members but women being over 80% in each. This means that the minimum expected beneficiaries from Jobs and Oils are 1600 people (that is, 20 members X 80 enterprises /groups). This translates into 1280 women and 320 men, all these being of youth age group who will support or benefit O&G exploration and production activities in the Albertine or in Uganda as a country and beyond.

Source: Living Earth Uganda, Project Report (2018)

areas with a view to supporting poverty reduction and sustainable development and contributing to the generation of new employment opportunities (see Table 10.4 below for a case study example).

Impact on the Environment

Oil and gas activities in the country can impact the environment from several angles. They may affect human beings, wildlife, and biodiversity, together with the associated tourism. Transboundary water resources and the economy of the Albertine Graben ecosystem where production is located are anticipated to bear the greatest threat. Furthermore, the corridor in which any pipelines or other transport systems will pass will also be affected. This policy considers environmental protection to include both the physical and social aspects and seeks to mitigate typical forms of environmental damage and hazards associated with oil and gas exploration, development, and production *(How effectively this has been done is a matter of field work).* The policy also supports gas utilization rather than wastage and therefore promotes prohibiting the venting of gas and discourages flaring of oil and gas except in situations of emergency where the alternatives may result into greater risk *(How effectively this has been done is a matter of field work)*. All these, if well implemented lead to mitigating environmental degradation, such as increased greenhouse effects arising from venting and flaring. Utilization of natural gas, especially in homesteads will contribute to reduction in the use of biomass energy which results in indoor house air pollution with undesirable effects on human health. The reduction in the use of biomass energy resulting from the increased use of natural gas will also have a significant impact on reducing the degradation of the environment through felling of trees. Access roads will be kept at a minimum in wildlife areas and other areas of sensitive biodiversity, and efforts will be made to construct and maintain these roads in a manner that conserves the environment.

This policy, in collaboration with other relevant policies, seeks to support control measures against the release of hazardous gases, chemical wastes and spills into the atmosphere, water bodies, aquifers, and soils which will ensure that water remains safe for animals, fish, and human consumption *(How effectively this has been done is a matter of field work)*. The policy supports relevant institutions to put in place disaster preparedness and response mechanisms for any oil spills in the Albertine Graben, storage facilities, and any transportation corridors. Putting in place these mechanisms together with their implementation shall be the responsibility of the Licensees. This policy endorses the principle of "polluter pays." Government is to design a mechanism of levying penalties for environmental pollution and/or degradation *(Whether this has been done, is a question that field work can answer)*.

Impact on Health

This policy recognizes several potential causes of negative impacts on human health from oil and gas activities. Oil spills can contaminate water sources leading to sickness and disease. Gas blowouts result in fires that destroy property and may lead to loss of human lives *(How prepared the country is to handle this is a matter of field work)*. Gas flares and dust result in air contamination leading to sickness. This policy promotes the setting up of best international practices for prevention and rapid emergency response mechanisms designed to mitigate against air and water pollution. Licensees are required to put in place the response mechanisms in all aspects of their operations. Efforts to construct roads in a manner that reduces or prevents dust pollution were meant to be promoted.

Additionally, this policy recognizes that drilling in settled communities and water bodies used by the population could be hazardous. Where deviation/directional drilling can minimize these hazards and achieve the desired results of the drilling objective in an efficient manner, deviation drilling was to be promoted. The policy also recognizes that health hazards do not occur in isolation from each other. While pollutants and toxins are directly inhaled by humans causing disease, they also invade the food chain entering fish, animals, and vegetables. Thus, the quality of water and food needs to be monitored and tested for unacceptable levels of pollutants and toxins such as lead and mercury. This policy, in collaboration with other relevant policies, supports the review, updating, and implementation of waste disposal standards together with the establishment and enforcement of the necessary monitoring, evaluation, and control mechanisms. Considerations were to be made as much as possible for the use of alternatives to toxic chemicals *(Whether this has been done, is a question that field work can answer)*. The policy also recognizes that the population movements and increased income could result in the increase in communicable diseases including HIV infection. The policy thus supports measures to mitigate the spread of these diseases *(How effectively this has been done is a matter of field work)*.

Impact on Education and Research Plans

Development of the oil and gas sector will create a demand for skills currently not supplied by the country's education and training institutions. This policy recognizes that under the Universities and Other Tertiary Institutions Act 2001, The National Council for Higher Education (NCHE) has the mandate to support the implementation of this policy through development and training of requisite national expertise and to promote relevant research in the oil and gas sector *(How effectively this has been done is a matter of field work)*. This policy promotes the establishment of human resources requirements for oil and gas activities and their projected estimates for the short, medium, and long term as well as the use of these requirements to support the training necessary to meet the demand with Ugandan nationals. Educational and training institutions in the country were to be induced to orient some of their training courses to meet human resource requirements of the oil and gas sector *(How effectively this has been done is a matter of field work)*.

Impact on Relations with Neighboring Countries

The Albertine Graben which is the principal area with the potential for petroleum production in the country is shared between Uganda and the Democratic Republic of Congo. An "Agreement of Cooperation for the Exploration of Hydrocarbons and Exploitation of Common Fields" was signed between the Government of the Republic of Uganda and The Executive Council of the Republic of then Zaire in 1990. This policy recognizes among other things, that on unitization, the method used in determining how oil and gas fields falling across borders or boundaries are shared, is well-established internationally, and used in several parts of the world. This policy supports the implementation of the Agreement of Cooperation, and the application of unitization as a basis for sharing any oil and gas fields falling across any borders. The policy also supports a review of the agreement of cooperation with a view to bringing it in line with the current and future requirements of the oil and gas sector *(How effectively this has been done is a matter of field work)*. In addition, the policy promotes transboundary cooperation based on existing foreign policy principles and agreements, and complements the programs, projects, and protocols formulated under the International Conference on the Great Lakes Region aimed at transforming the region from tensions, hostilities, and underdevelopment to security, stability, and prosperity.

There has been need for consultations with neighboring countries for access where transport corridors like pipelines may need to cross borders. In addition, sales of the country's oil and gas may also require discussions or entry into agreements with foreign countries. This policy supports, where necessary, the enhancement of the existing foreign policy principles with a view to facilitating harmonious implementation of these activities.

Impact on the National Economy

This policy shall, in collaboration with other relevant national policies, promote the use of oil and gas activities to have a positive impact on the economy leading to the generation of revenues for investment in the non-oil and gas sectors resulting in higher growth rates of the economy.

Conclusions

Most O&G companies (upstream and downstream) operating in Uganda are transnational enterprises. They are signatories to international bodies (like WBCSD, EITI, OECD, OPEC, and UN Global Compact), which compel them to voluntarily act in a socially responsible way (that is CSR/Social Compliance) in addition to meeting mandatory legal requirements in the geographic locality of their operations. This compelling requirement for O&G players to engage in CSR/Social Compliance, is in fact explicitly mentioned by "The National Oil and Gas Policy for Uganda (2008)," objectives 9(c)[3] and 10(f).[4] However, Uganda being a developing country, the O&G players bypass this requirement given its voluntary nature, and they majorly concentrate on provision of EIA/ESIA purposely to secure a "legal license to operate" from regulators (NEMA and Petroleum Authority of Uganda—PAU). Additionally, supervision to check on the CSR compliance as promised in the EIA/ESIA is very weak. This is because neither the government of Uganda nor O&G companies have internal controls to monitor the impact of their CSR (Katamba & Wickert, 2021).[5] For example, a scrutiny of the CNOOC's[6] Sustainability reports featuring Uganda has shown that most of the CSR reporting is not prepared with a reality check and misses critical key GRI/ISO 26000 standard information. In fact, the CSR reporting is not prepared through "third party controls/reality check" since

[3] Objective 9(c) Requires oil companies and their contractors/subcontractors to use self-regulation and best practices in ensuring environmental protection and biodiversity conservation.

[4] Objective 10(f) Encourage the implementation of corporate social responsibility by oil companies.

[5] Katamba D., and Wickert C.M.J. (2021) Corporate Social Responsibility in Uganda. In: Idowu S.O. (eds) Current Global Practices of Corporate Social Responsibility. CSR, Sustainability, Ethics & Governance. Springer, Cham. https://doi.org/10.1007/978-3-030-68386-3_26

[6] CNOOC (2017), Environmental, Social & Governance Report, https://www.cnoocltd.com/attach/0/1f427f28d10d457aa1752c2206cc91cb.pdf, accessed October 3rd, 2021.

most activities, do not translate into tangible action (development impact) in Uganda. The NPD-3 has attempted to explain this discrepancy [in "CSR reported Vs CSR Reality on ground"] by stating that:

In addition to updating academic knowledge in Energy Economics, Uganda's Ministry of Energy and Mineral Development (MoEMD), Petroleum Authority of Uganda (PAU) and Uganda National Oil Company (UNOC) must devise policy shifts in the Gas and Oil subsector. Specifically, demonstrating a need to shift from overdependence on Environmental Impact Assessment (EIA) as the only major pre-requisite to be granted a license to operate in Uganda's O&G subsector, to also demanding comprehensive and operationally relevant integrated Corporate Social Responsibility (CSR) plans which deliver Social Compliance[7] in support of realizing Uganda's National Development Priorities as enshrined in NDP 3 [see Priorities[8],[9]], Vision 2040 as well as SDGs (12, 8, 7, 5, 1). Such a requirement will help downstream and upstream O&G players to assess and profile CSR activities that engage men, women, and youths in efforts promoting sustainable social, economic, ecological development, environmental responsibility, and climate change resilience.

Additionally, for Local Governments of in the Albertine districts to include Buliisa, Kikuube and Hoima districts, and Ministry of Gender, Labor, and Social Development (MoGLSD), need to have their capacity development in the context of how they can:

(i) monitor the relevance of O&G players' CSR to policy instruments that promote social compliance & SDGs.
(ii) make informed decisions about the readiness of private companies to enter Memorandum of Understanding (MoU) for potential community development PPP projects; and
(iii) make strategic plans for social development & gender mainstreaming.

[7] Social compliance is a continuing process in which organizations endeavor to protect the health, safety, and rights of their employees, the community and environment in which they operate, and the lives and communities of workers in their supply and distribution chains.

[8] NDP 3 priority of *"Community Mobilization and Mindset Programme"* aims to empower families, communities, and citizens to embrace national values and actively participate in sustainable development. Key expected results include increased participation of families, communities, and citizens in development initiatives; enhanced media coverage of national programs; increased household savings; increased social cohesion and civic competence; and better uptake and/or utilization of public services (education, health, child protection, etc.) at the community and district level.

[9] NDP 3 priority of "*Natural Resources, Environment, Climate Change, Land and Water Management Programme*" aims to stop, reduce, & reverse environmental degradation and the adverse effects of climate change as well as improve utilization of natural resources for sustainable economic growth and livelihood security. Key expected results include improved land use and management; increasing land area covered under forests and wetlands, increasing compliance of water permit holders with permit conditions, and enhancing the accuracy of meteorological information.

Considering the above conclusions, we suggest these specific action points for Petroleum Authority of Uganda (PAU) which we think can safeguard Uganda against the 'Oil Curse.'

(i) Devise policy shifts from over dependence on Environmental Impact Assessment (EIA), to also demanding for comprehensive & operationally relevant integrated CSR plans which are responsive to Social Compliance, as depicted in NDP 3 (or as amended), and Vision 2040.
(ii) Develop tools that can be used to monitor the relevance of O&G players to policy instruments that promote SDGs so that the O&G subsector can also direct participate in accelerating Uganda realize its commitment to SDGs framework.
(iii) Develop a template which can be used by Local Gov'ts or women and Youths groups in assessing the readiness of O&G sector players before entering Memorandum of Understanding (MoU) for projects in the Albertine.
(iv) Design guides to help local communities/councils in O&G mining areas to make CSR and Social Compliance strategic plans for their communities which they can present to O&G players.

References

Brundtland Commission. (1987). *Report of the world commission on environment and development: Our common future*. Annex to UN General Assembly document A/42/427.

Business Call to Action. (2013). *Innovations that improve lives*. Accessed Mar 26, 2021, from http://www.businesscalltoaction.org/

CNOOC (2017). *Environmental, social & governance report*. Accessed Oct 3, 2021, from https://www.cnoocltd.com/attach/0/1f427f28d10d457aa1752c2206cc91cb.pdf

Katamba, D., & Wickert, C. M. J. (2021). Corporate social responsibility in Uganda. In S. O. Idowu (Ed.), *Current global practices of corporate social responsibility*. CSR, Sustainability, Ethics & Governance. Springer. https://doi.org/10.1007/978-3-030-68386-3_26

Katamba, D. (2018), *Shaping and Managing Corporate Social Responsibility in a Low-Income Country: Lessons from Uganda*, PhD Thesis, Accessed Mar 26, 2021, from http://repository.londonmet.ac.uk/2940/1/KatambaDavid%20-%20Final%20Hardbound%20Thesis.pdf

Katamba, D., Nkiko, C., Tushabomwe-Kazooba, C., Babiiha, M., & Kemeza, I. (2014). Community involvement and development: An intermarriage of ISO 26000 and millennium development goals. *International Journal of Social Economics, 41*(9), 837–861.

Kuteesa, A. (2014), Local communities and oil discoveries: A study in Uganda's Albertine graben region, economic policy research Centre (EPRC).

Long, B. S. (2022). CSR and reconciliation with indigenous peoples in Canada. *Critical Perspectives on International Business, 18*(1), 15–30. https://doi.org/10.1108/cpoib-12-2017-0096

Mwakali, J. A., & Byaruhanga, J. N. (2018, December). *Human resources and oil in Uganda: An analysis of Uganda's human resource development for the oil sector*. Centre for Research on Peace and Development. https://soc.kuleuven.be/crpd/workingpapers

Ross, M. L. (2012). *The oil curse: How petroleum wealth shapes the development of nations*. Princeton University Press.

Rwengabo, S. (2017). *Efficiency, sustainability, and exit strategy in the oil and gas sector: Lessons from Ecuador for Uganda, ACODE policy research paper series no. 81*. ACODE.

Ministry of Energy and Mineral Development. (2008). *National oil and gas policy for Uganda*. Policy paper. Ministry of Energy and Mineral Development.

David Katamba, PhD, is an Associate Professor of Corporate Sustainability and Responsibility. I am an expert evaluator with Economic and Social Research Council (ESRC), a part of UK Research and Innovation (UKRI). I initiated and led the consultative process that established the United Nations (UN) Global Compact Network in Uganda, which is currently hosted by Federation of Uganda Employers (FUE). I am a World Bank Institute trained Corporate Sustainability and Responsibility (CSR) professional with over 16 years' experience in Higher Education quality management systems, sustainable development aspects including in youths and women entrepreneurship, Education, and Business. Currently (Jan 2023–June 2026), I am coordinating a € 1,000,000 EU Erasmus+ funded project, 'Strengthening Capacity for ICT Usage and Policy Reforms for Relevant and Quality Masters Research Process Management in Uganda's Higher Education Institutions (ICT-4MRPQ)', (project No. 101083105). He is a certified CSR Consultant/Advisor and trainer for International Business Leaders Forum (IBLF) and Aga Khan Development Network (AKDN). I have won and led implementation of several international research grants including from €300,000 Netherlands' NWO WOTRO to implement AGRI-QUEST project (Project No. W 08.270.317). I was also part of team that won €801,000 from EU Erasmus+ [call, EAC/A03/2016, research project No: 574218-EPP-1-2016-1-NO-EPPKA2-CBHE-JP to implement CapSEM project (2016-2019). Dr Katamba has supported several HEIs in and outside Uganda to strengthen their academic quality processes through coaching services, matchmaking, financial resources mobilization, and capacity building. I have been (and I am still) a peer reviewer of proposals, grant applications, publications for several academic journals, and projects for several development organizations including for European Commission's EACEA. Additionally, he acted as mentor, coach, and match makers for 'Start Your Business (SYB)' scheme and competitions by World Bank and Private Sector Foundation Uganda (PSFU) as well as other SME and trade and investment initiatives. He has delivered services and implemented projects in Netherlands, Belgium, Canada, Turkey, Italy, and East Africa. Lastly, he helped to start the United Nations (UN) Global Compact Network in Uganda.

Bosco Amerit (BCom, MBA, LLB, PDLP, LDC) lectures at Makerere University Business School (MUBS) and coordinates the Norwegian Programme for Capacity Building in Higher Education and Research for Development at MUBS; 2016–2021 NORHED I and 2021–2026 NORHED II being implemented at MUBS in partnership with the Norwegian University of Life Science (NMBU) and the University of Dar es Salaam. He participated in the implementation of various projects including but not limited to the EU-funded CapSEM project, the Good Governance Mentorship Project, under the Resilience Innovation Challenge for Conflict (RIC4COF) 2016, supported by the East Africa Resilience Innovation Lab and the US Global Development Lab powered by USAID and three research and innovation projects of Makerere University, supported by the Government of Uganda. He is currently pursuing doctoral studies in energy economics and governance and holds three degrees: Master of Business Administration from Makerere University (2010), Bachelor of Commerce (accounting) degree from Makerere University (2003), Bachelor of Laws (LLB) and a Postgraduate Diploma in Legal Practice from the Law Development Centre, Uganda. He is enrolled as an Advocate and teaches management courses including business administration, management theory and application, strategic management, and business communication. Lastly, as a lecturer/scholar of business and management/lawyer and a finalizing PhD study in energy economics and governance, his research interest is in the green fuels, energy governance, and sustainability. During and after the doctoral studies, Bosco intends to dedicate a greater part of his research career to conducting applied research as well as developing models that seek to inform policy and decision making within the energy sector.

Maureen Basuuta is a Lecturer in the Department of Economics at Makerere University Business School (MUBS). Her research interests are in poverty, informal sector, and the energy sector particularly energy poverty and the involvement of women in the oil and gas industry. She has worked on different research projects such as the Makerere University Research and Innovations (RIF) funded projects and the NORAD and NORHED Research Grants for Capacity Building in Education and Research for Economic Governance in Uganda. She has presented several papers at different conferences and published in sustainable accounting. Maureen is also a consultant and has consulted for different organizations such as Joint Medical Store, UN Women, and the Law Development Centre in Uganda. Maureen is also a trainer for several capacity building training programmes.

Swithern B. Tumwine, Executive Director of Living Earth Uganda (LEU), is a seasoned, innovative, and experienced manager with a wealth of knowledge of over 25 years in institutions'/projects' leadership, management, and governance. These institutions include Financial Institutions and NGOs where his focus has been in areas of banking, microfinance, financial management, marketing, and linkages of entrepreneurs/farmers with financial institutions, CSR issues, and guidance in sustainable productivity of such entrepreneurs, institutional planning, monitoring, evaluation, community mobilization for awareness and mobilization, and general management. He has recently also focused on livelihood restoration programmes for project affected persons/communities in oil and gas communities in Uganda where CSR education and practice is being promoted among other issues discussed with international oil companies. He holds a master's degree in financial management from Uganda Management Institute (UMI), Postgraduate diploma in Financial Management (UMI), Bachelor's degree in Commerce from Makerere University, Kampala, Uganda, and certificates in computer applications. He has undertaken various training both locally and internationally that include training of trainers in community livelihood training techniques by Living Earth Foundation, London, UK (2010), financial institutions systems development and leadership by Opportunity International Leadership Centre, San Diego, California, U.S. (2004), Microfinance Training in Best Practices Certificate, Naropa University, Boulder, Colorado, U.S. (2002), Management Information Systems training, Opportunity International Center, Denver, Colorado, U.S. (2001), and Monitoring and Evaluation in Small and Micro Enterprise Development, the Springfield Centre for Business Development, Durham, UK (1999) among others. He has worked as Executive Director for Living Earth Uganda (2007 to-date). Previously held positions of Executive Director, Uganda Agency For Development (UGAFODE) Ltd (1997-2005); Assistant Director, Presidential Initiative on Banana Industrial Development in Uganda, Manager, Uganda Commercial Bank (UCB) (1994-1996); Banking Officer, UCB (June 1989-1993); Assistant Accountant, Continental Building Society (July 1988-June 1989).

Muhammed Ngoma, PhD, is a Professor of Business Administration and Management. He holds a bachelor's degree in commerce, a Master's and a PhD in Business Administration. He is currently the Chairman Board of Directors of The Uganda Security Printing Company, the State-Owned Enterprise mandated to produce/supply Security documents and their corresponding IT support systems to all government MDAs (including driving permits, passports, National IDs, Birth and Death Certificates among others). Prof. Ngoma is the outgoing Dean of the Faculty of Graduate Studies and Research at Makerere University Business School, and currently the Director of The Energy Economics Resource Centre at the university. Prof. Ngoma has held several leadership and management positions both within and outside the university for the last 18 years. Prof. Ngoma Muhammed has over 60 publications in refereed journals and has supervised over 100 PhDs and masters' students to completion. He has won several grants for this country and has consulted for several organizations ranging from the public to the private sector. He sits on several boards and councils and is a member of the Phi-Beta-Delta Honor Society of Scholars and Researchers. Prof. Ngoma was recently ranked among the top scientists in the country among several other awards/rankings. Prof. Ngoma Muhammed is married, with six children (3 boys and 3 girls). He enjoys reading, travelling, basketball, volleyball, dancing, and helping the vulnerable people.

Part III
Corporate Social Responsibility and SMEs

CSR in SMEs in a Sub-Saharan Africa: Motivations, Perceptions, Practices, and Barriers to Implementation
John O. Okpara and Jet Mboga

Responsible Entrepreneurship: Raising Corporate Social Responsibility Awareness Among Male and Female Owned and Managed Small and Medium Sized Enterprises (SME's) in Abeokuta, Ogun State Nigeria
Adebimpe Adesua Lincoln, Brendhain Diamond, and Jane Croad

Chapter 11
CSR in SMEs in a Sub-Saharan Africa: Motivations, Perceptions, Practices, and Barriers to Implementation

John O. Okpara and Jet Mboga

Introduction

Small and medium-sized enterprises (SMEs) are the main driving force in Nigerian economic development and also represent a key component of the Nigerian economy. However, despite their undeniable importance, studies on SME CSR remain rare. In much of CSR discourse, there is often a misconception that CSR activities are carried out only by large and highly profitable businesses (Jansson et al., 2017; Lucky, 2018; Meyer et al., 2017). SMEs comprise the growth engine of many economies in the word, contributing to their development, job creation, gross domestic product (GDP), gross national product (GNP), and exports (PwCs, 2022). Although it is relatively easy to identity CSR practices in large companies because of their publicity, the same does not happen at the SME level (Perrini et al., 2007; Tilley, 2000). Globalization of business has put SMEs under immense pressure to compete for survival. Consequently, SMEs need to find new avenues for survival and expansion. At the same time, they also feel the pressures from stakeholders who demand their interests to get served. Such an environment of intense competition has demanded that private enterprises respond to the conflicting needs of stakeholders in a more proactive and sincere way.

A review of the literature on subject shows that a great deal of research has been done on CSR and SMEs. However, these studies have been primarily conducted in the West. Little research on this topic has been conducted in Africa, particularly in sub-Saharan Africa. This creates a knowledge gap in our understanding of this

J. O. Okpara (✉) · J. Mboga
Department of Management, Zeigler College of Business,
Commonwealth University of Pennsylvania, Bloomsburg, PA, USA
e-mail: jokpara@commonwealthu.edu; jokpara@commonwealthu.edu

U. Nwagbara et al. (eds.), *Corporate Social Responsibility Disclosure in Developing and Emerging Economies*, CSR, Sustainability, Ethics & Governance, https://doi.org/10.1007/978-3-031-61976-2_11

important topic. The purpose of this research is to bridge that knowledge gap to increase our understanding on the role of SMEs in promoting CSR in Nigeria.

Given the highly specific understanding of the CSR concept and practices by SMEs, and given the knowledge gap that exists in our understanding of this subject in Nigeria, it is clear there is a general lack of awareness of the most common CSR practices, motivations, and existing obstacles in the country. This paper examines the motivations, perceptions, and barriers hindering CSR practices at SMEs in Nigeria. This study is an important stepping stone toward in-depth understanding of Nigerian SMEs' CSR practices, motivations, and barriers to effective implementation of CSR initiatives.

The rest of the chapter is structured as follows: The next section on literature review discusses prior studies on CSR and SMEs. This is followed by a methodology section, which discusses the methods and processes used in conducting this study. The subsequent section discusses the results of this study and also highlights directions for future research based on these findings. Finally, a section that discusses the limitations of the study and makes suggestions based on the identified limitations concludes the paper.

Literature Review

The Meaning of Corporate Social Responsibility

There are several definitions of SMEs available in the literature (Ahmad et al., 2021; Olusegun, 2012); the most common definition comes from the Organization for Economic Cooperation and Development (OECD) in the context of the European Union (EU). OECD defines an SME as an enterprise that has no more than 250 employees with financial assets not exceeding EUR 50 million, whereas an enterprise with a workforce between 10 and 49 and a financial asset of less than EUR ten million (OECD, 2005). In the context of Nigeria, SMEs are enterprises with a maximum asset base of 500 million Naira, excluding land and working capital, and with a staff strength of not less than 10 and not more than 300. There are many definitions of CSR but a lack of consensus on one (Dahlsrud, 2008). The lack of a generally acceptable definition of CSR means that the owners or founders of a firm have different perceptions of social responsibility. Therefore, a firm's social responsibility depending on the perception of the individuals who make the decisions gives rise to our first research question.

RQ1: What are the CSR perceptions of SME owners and managers in Nigeria?

Motivations for CSR

Several studies have found that factors such as long-term financial strategy, competitive advantage, moral leadership, good corporate citizenship, image, avoidance of regulatory action, and pressure from stakeholders were the motivations most mentioned as reasons for firms to engage in CSR (Brønn & Vidaver-Cohen, 2009; Lynes & Andrachuk, 2008; Nejati & Amran, 2009). Motivations can be separated into two categories: intrinsic and extrinsic (Ryan & Deci, 2000). SMEs owners are intrinsically motivated to engage in CSR because they derive inner satisfaction from it (Ryan & Deci, 2000). Extrinsic motivation is perceived as the belief that CSR engagement leads to a separable positive outcome for the firm or, in other words, that there are external rewards for CSR (Ryan & Deci, 2000). Graafland and van de Ven (2006) found that although extrinsic, or strategic, and intrinsic motives were related, intrinsic motives had a stronger correlation with SME involvement in CSR. An inquiry into managers' motivations for CSR is thus valuable because it enables a better appreciation of why firms act the way they do. This leads to our formulation of RQ2:

RQ2. What motivates SMEs to engage in CSR activities?

CSR Activities

According to Fernando (2022), CSR initiatives are often broken down into four categories: environmental, philanthropic, ethical, and economic responsibility. Environmental initiatives focus on preservation of natural resources, whereas philanthropic initiatives focus on donating to worthy causes that may not relate to a business. Ethical responsibility ensures fair and honest business operations, whereas economic responsibility promotes the fiscal support of the goals above (Fernando, 2022). According to a *Business Day* (2014) report, most SMEs in Nigeria embark on community responsibility projects, and when they do, they avoid the term *CSR*, preferring instead donation, philanthropy, charity, contribution, or gift. Therefore, Nigerian SMEs mostly engage in donations, philanthropy, charities, contributions, or gifts. When SMEs participate in philanthropy activities and give back to communities, they believe that they have fulfilled their social obligations, which to them is CSR without using the specific term. Awards and recognitions for CSR initiatives have mostly been given to large corporations, not to SMEs. This does not mean that they do not engage in giving back because their concept and application of CSR is narrowed. Based on the above analysis, we therefore propose RQ3:

RQ3: What CSR activities do SMEs in Nigeria engage in?

Barriers to SMEs Participation in CSR

One of the most important aspects of CSR in SMEs in developing countries relates to the barriers they face in adopting CSR practices. The CSR initiatives of SMEs in developing countries tend to be hindered by several barriers that generally prevent them from engaging in CSR activities. Some of the barriers are discussed in the following sections.

Lack of Management Commitment to CSR

CSR is a significant issue and should be given a serious consideration by the senior management of a company (Reimer et al., 2018). Top management's commitment affects the strategic intent of a business toward its social responsibility (Goyal & Kumar, 2017). Therefore, senior management's commitment can encourage others to ensure the implementation of effective CSR (Yusliza et al., 2019). Whatever the managers do—in which direction they push and how hard they push—dictates where they are going and their intent to fulfill their social responsibility. It follows that lack of senior management commitment to CSR has been acknowledged as a critical barrier to implementing a CSR program (Zou et al., 2021). More specifically, top management's commitment provides a direction for CSR and is essential for implementing a CSR plan in their organizations. However, lack of top management commitment is usually regarded as a critical barrier for CSR implementation. One can therefore surmise that lack of top management commitment to CSR is a barrier to implementation of CSR in SMEs. This leads us to propose RQ4:

RQ4. Is lack of management commitment a barrier to SMEs' CSR implementation?

Lack of Financial Resources

Lack of financial resources restricts most SMEs from using funds to invest in CSR-related activities (Mahmood et al., 2021). Lack of finance has been described as the main obstacle for most SMEs in developing countries to implement CSR. According to Goyal and Kumar (2017), lack of money and consumers' passive attitudes toward CSR are the major barriers in CSR implementation. Although the concept of CSR is gaining attention in developing economies, its adoption by corporations is still in the formation stage. Generally, organizations in developing countries consider CSR as an expenditure rather than a business opportunity (Mahmood et al., 2021). According to Mahmood et al. (2021), to date, there is no universal consensus in recognizing CSR as a cost-effective strategy, implying that some companies, especially those is developing regions, do not allocate a dedicated financial resource for CSR initiatives. Therefore, lack of financial resources is also something that has been recognized in the extant literature as a barrier to CSR implementation. We, therefore, propose the following research question:

RQ5. To what degree is the lack of finance a barrier to SMEs CSR implementation?

Lack of Understanding About CSR

Although the concept of CSR has been in the discussion for more than three decades, no unanimous definition of CSR exists. This is the main reason why the concept has been mixed with other areas such as business ethics, corporate citizenship, corporate governance, and others beyond the interests of companies and the requirements of the law. There are different definitions of CSR, but the definition by Carroll (1991) is one of the most cited. According to Carroll (1991), CSR is the economic, legal, ethical, and philanthropic obligation of a business toward all stakeholders. Researchers have indicated that the concept of CSR is an ever-evolving one (Gupta et al., 2021; Singh & Misra, 2022), and because of that there exists a lack of understanding of the concept. More specifically, it has been reported that CSR has not received a clear explanation, at least from the perspectives of employees and corporate management in developing countries (Goyal & Kumar, 2017; Kumar et al., 2019). Several researchers have stated that lack of understanding about CSR is a critical barrier that hinders an SME's adoption of CSR (Merwe & Wocke, 2007; Shen et al., 2015). From the forgoing analysis, we propose RQ6.

RQ6. Is a lack of understanding a barrier to CSR implementation?

Lack of Strategic Planning

An effective strategic planning process is expected to place CSR close to the business operations (Lam & Lim, 2016). According to Freeman and Moutchnik (2013), a well-planned strategy enables companies to identify, understand, and respond effectively to CSR implementation. Lack of effective strategic planning for CSR can create obstacles in CSR implementation. To achieve the full flux of the concept, CSR should be an integral part of strategic planning. Without a comprehensive strategic plan for CSR, a corporation may find it difficult to execute a CSR plan efficiently (Sitnikov & Bocean, 2017). Therefore, the lack of strategic planning for CSR is considered a critical barrier for CSR implementation. Therefore, we propose RQ7:

RQ7. Is a lack of strategic planning for CSR a barrier to CSR implementation?

Complexity of CSR

CSR is a wide-ranging concept that involves multiple facets that interact with different stakeholders. Therefore, SMEs must learn how to carefully interact with stakeholders with diverse interests. CSR initiatives and potential business outcomes vary by business, type of sector, and range of companies. CSR implementation requires careful consideration between business and society; companies may find it difficult to implement to achieve the desired outcomes. Several researchers have indicated that CSR is a complex concept that is difficult to implement (Arevalo & Aravind, 2011; Bux et al., 2020; Hsu & Cheng, 2012). Therefore, this complexity is a critical barrier hindering CSR implementation. Thus, we propose RQ8.

RQ8. Is the complexity of CSR a barrier to SMEs' CSR implementation?

Research Methodology

The research design involved administering a questionnaire to randomly selected SMEs. The data were collected from December 1, 2021 to March 1, 2022, using a questionnaire survey method. The sample comprised firms listed in the Nigerian Business Directory, published by the Manufacturers Association of Nigeria (MAN).

Sample and Procedures

A random selection of 110 firms from the Nigerian Business Directory was contacted via letter, email, and text requesting their participation in the research. The firms identified from the directory was personally contacted by telephone, email, or text for permission to participate in the study. The participating SMEs were in Aba, Abuja, Kaduna, Lagos, and Onitsha in Nigeria. These cities were selected because they were considered to have the largest concentration of SMEs and would facilitate the data distribution and collection process. To avoid losses and delays due to unpredictable communication systems in Nigeria, a drop-off and pick-up method was used. This method ensured reliable distribution and collection procedures, which were systematic and controlled by the researchers. Eight trained research assistants and four field supervisors were responsible for the distribution and collection of the survey questionnaires. They gave the questionnaires to a contact person in each of the firms identified for the survey along with a personally addressed cover letter. This contact person was requested to distribute the questionnaires to the selected managers or owners and then affix to a central place the large envelope provided. The identified people were asked to anonymously complete the questionnaire and deposit it into the large envelope. The contact person was asked to seal the envelope and have it ready to be collected in 2 weeks. Researchers contacted each contact person by telephone, email, or text within 4 days of the return deadline as a reminder that someone would be coming to pick up the envelope. The survey instrument used in this study was based on measures used previously in the CSR literature (Carroll, 1991).

Interview Method

In the second stage of data collection, we conducted in-depth telephone interviews with 10 SME owners and managers to seek insights on their CSR, motivations, practices, and barriers. This type of interview with those who are influential, prominent, and well informed in their fields is known as an elite interview. In these elite interviews, interviewees are not subject to standardized questioning but are allowed to introduce "their notions of what they regard as relevant, instead of relying upon the investigator's notions of relevance." We found this approach useful because these elites were able to elaborate on the position of CSR in their companies.

Table 11.1 shows sample characteristics for the study. A total of 110 questionnaires were distributed to the owners and managers of SMEs. Of the 110 SMEs surveyed, 100 were returned, 88 of which were usable for analysis, yielding a response rate of 88%. Of the 88 respondents, 20 (22%) were between the ages of 20 and 40, 42 (48%) were 41–52, and 26 (30%) were 43–62 years old. The overwhelming majority of the respondents were male (65%) and 35% were female. With respect to experience, 28% had about 1 year of experience, 34% had 2–3 years' experience, and 38% had more than 4 years. In terms of education, 34% had a bachelor's degree; 30% had high school diploma; 25% had either OND, HND, or NCE diplomas; and 11% had a master's degree. Most of the respondents were owners of the firm: 45% were owners, 41%, were managers, and 14% were supervisors.

Data were collected from the following cities: Aba 23%, Abuja 22%, Kaduna 14%, Lagos 21%, and Onitsha, 22%. Table 11.2 shows the number of firms selected for the study and the companies represented in the survey. The sample represented diverse companies; automobile dealers (8%), automobile parts dealers (7%), building material businesses (11%), roadside fast food restaurants (9%), hotels (14%), Internet cafés (7%), motorcycle (10%), motorcycle parts dealers (9%), restaurants (11%), and supermarkets (14%). All of the companies surveyed (100%) served the local markets.

Table 11.1 Demographic variables of the sample ($n = 88$)

Demographic variables	Number	Percent
Age		
20–40	20	22
41–52	42	48
53–62	26	30
Gender		
Male	57	65
Female	31	35
Experience		
0–1 year	25	28
2–3 years	30	34
4–5 years	33	38
Education		
Doctoral degree	0	0
Master's degree	10	11
Bachelor's degree	30	34
OND, HND, NCE	22	25
High School Diploma	2	30
Position		
Owners	40	45
Managers	36	41
Supervisors	12	14
Location (Cities)		
Aba	20	23
Abuja	19	22
Kaduna	12	14
Lagos	18	21
Onitsha	19	22

Table 11.2 Companies surveyed for the study

Type of business	Ownership	Market	Number	Percent
Automobile dealers	Private	Local	7	8
Automobile parts dealers	Private	Local	6	7
Building material businesses	Private	Local	10	11
Roadside fast food restaurants	Private	Local	8	9
Hotels	Private	Local	12	14
Internet cafés	Private	Local	6	7
Motorcycle	Private	Local	9	10
Motorcycle parts dealers	Private	Local	8	9
Restaurants	Private	Local	10	11
Supermarkets	Private	Local	12	14
Total			88	100

Results and Discussion

Perceptions of CSR

Table 11.3 shows the means and standard deviations of perceived factors of CSR. The most important items characterized by the highest means and the lowest standard deviations were environment ($M = 4.85$, SD = 0.30), and human rights ($M = 4.84$, SD = 0.40), whereas the only item with a mean lower than 4 was profitability ($M = 3.77$, SD = 0.81). The interviewees expressed mixed opinions toward profitability. A few interviewees did not think that being profitable is a part of a corporation's social responsibility. One interviewee commented that "being profitable is the basic requirement of any company that exists in the economy." According to this interviewee, it would be too "narrow-minded" to consider profitability as being part of CSR. Some other interviewees, however, felt that profitability is an essential part of CSR. More than half (57%) of the owners felt that CSR should be recognized as a core business function, whereas almost three-quarters (74%) agreed that social responsibility should be a consideration when formulating corporate strategies. *One interviewee felt that his corporation has to be profitable "to run the company efficiently—improve the system ... be responsible to his shareholders ... employees ... contribute to the community."*

Some interviewees felt that CSR should "come from the heart" rather than merely complying with government regulations or legislations. One manager commented that "CSR is not about what the government sees as right or wrong... . It's not what the government should do but what companies should do... . It is not about just operating within the rules." An interesting issue that emerged during the interviews was the balance between profitability and CSR and whether corporations need to be profitable before engaging in CSR actions. Table 11.4 shows the level of agreement or disagreement on the concept of CSR. As shown in Table 11.4, 81.5% agreed or strongly agreed that CSR should be a core business function, and 99.5%

Table 11.3 Means and standard deviations of SMEs' perceptions of CSR ($n = 88$)

Item	M	SD
Compliance with laws	4.22	0.70
Business ethics	4.30	0.50
Profitability	3.77	0.81
Corporate governance	4.15	0.87
Accountability	4.00	0.91
Environment	4.85	0.30
Health and safety	4.78	0.50
Labor practices	4.58	0.55
Human rights	4.84	0.40
Charitable contributions	4.52	0.88
Community involvement	4.44	0.77
Sustainability	4.50	0.64

Table 11.4 Percentages indicating the extent companies believe the following statements describe the concept of CSR ($n = 88$)

Statement	Strongly disagree (%)	Disagree (%)	Neutral (%)	Agree (%)	Strongly agree (%)
CSR is a publicity/image issue	18.0	24.3	17.2	15.0	30.5
CSR is making charitable contributions	0	4	4.7	45.2	46.1
CSR does not have a positive impact on financial performance	25.1	36.1	22.1	11.3	5.4
CSR should be recognized as a core business function	0	0	5	48.6	50.9
CSR should be a consideration while setting corporate strategy	0	0	3.0	49.7	47.3
Companies need not be concerned with society	2.1	2.2	1.5	48.0	46.2
CSR should be completely voluntary; no laws should govern it	43.3	48.5	3.2	2.8	2.2

agreed or strongly agreed that "CSR should be a consideration while setting corporate strategy" 97%. Table 11.3 also indicated that 91.3% of the respondents stated that CSR is about making charitable contributions. An overwhelming majority of the respondents, 91.8%, stated that CSR should be completely voluntary and that no laws should govern CSR activities. The majority of the respondents (100%) agreed that the items listed in Table 11.3 should make an SME socially responsible. One respondent stated that *"to me the most important dimensions of CSR are making charitable contributions [to] help the poor and community involvement in building and marinating local hospitals and health centers because [the] government absent in these areas." Another owner said that "as far as I'm concerned, the most important aspect of the CSR is environmental issues, and companies should do more and must take environmental issues more seriously."*

Motivations for CSR

With respect to motives, our findings show that the top three motivators as shown in Table 11.5 were enhance reputation (M = 4.80, SD = 0.90), attract investors (M = 4.53, SD = 0.70), and enhance community trust and support (M = 4.80, SD = 0.90). Other motivators for practicing CSR include increasing profits (90%), long-term sustainability (96%), external pressures (91%), altruism (90%), and recruiting and retaining employees (94%). It is important to note that more than half of the sample did not consider government regulation as a motivational factor.

Table 11.5 Motivations for practicing CSR ($n = 88$)

Variable	*M*	SD	%
Increasing profits	3.68	0.78	90
Long-term sustainability	4.35	0.86	96
Company tradition	3.80	0.83	92
Recruiting and retaining employees	4.02	0.77	94
Attracting investors	4.53	0.70	95
Promoting transactions/partnerships	3.87	0.91	89
Enhancing community trust and support	4.80	0.90	97
Avoiding regulations	2.90	0.92	88
Favorable media coverage	3.6 0	0.88	86
Improving public welfare	4.40	0.89	98
Altruism	3.88	0.78	90
External pressures (NGOs, consumers, community, etc.)	3.87	0.80	91

Table 11.6 Activities undertaken by SMEs in the sample ($n = 88$)

Activity	Frequency	Percent
Community projects	81	92
Educational projects	79	90
Sporting events	75	85
Health care projects	71	81
Charitable donations	84	85
Volunteer activities	80	91
Training and development	55	62
Employee welfare initiatives	57	65
Environmental projects	51	58
Ethics codes	18	20

CSR Practices

Table 11.6 shows activities undertaken by the SMEs surveyed in this study. Respondents carried out the following activities: ethical codes of conduct (20%), environmental projects (58%), training and development (62%), employee welfare initiatives (65%), sporting events (85%), charitable donations (85%), health care projects (81%), educational projects (90%), volunteer activities (91%), and community projects (91%). The interviewees commented that SMEs implement socially responsible activities either for practical or moral reasons. One of the survey respondents added that in his company they implement managerial and operational systems to obtain certifications and to have access to market opportunities. Another executive commented as follows: *"If we [SMEs] do not do something, then we are all guilty of what the world is currently facing… . This is why I consider I need to have an impact on my circle of influence. I know [it] is not a great thing, but if I do not do this, I would not be able to operate."*

Barriers to CSR Implementation

As shown in Table 11.7, many of the SMEs surveyed agreed that lack of financial resources (92%), lack of a strategic plan for CSR (90%), lack of government support for CSR (88%), the complexity of CSR to implement (88%), lack of regulations for CSR (84%), lack of knowledge about CSR (85%), and lack of management commitment (80%). According to the respondents, these are the critical barriers that hinder the SMEs in Nigeria from practicing CSR. For example, the lack of resources restricts most SMEs from sparing funds to be invested in CSR-related activities. They are unable to implement CSR because they do not fully understand the concepts and believe it is costly, which will affect the cash flow of the company, which suggests that financial constraints limit the capability of the company to commit to CSR practices. Because they do not understand the concept, the majority of respondents requires a longer time and needs systematic guidance and training for them to implement CSR. Here is summary of what interviewees said about barriers to implementation:

> Regarding financial constraints, one of the respondents stated, *"The fact is, our organization wants to go engage in CSR activities, but we do not have sufficient funds for that. We do not receive any financial help (interest-free or low-interest loans) implement sustainable practices."*
>
> On the knowledge barrier, one of the owners indicated, *"So far, the general perception here is that CSR is all about donations and charity, etc. I think most of us did not know that CSR is more than that."* Another executive stated, *"In my view, CSR is about donations to help the poor people in the community."*
>
> Concerning management commitment, an owner said, *"Our top management pays very little or no attention towards CSR initiatives. I do not remember even a single case during my stay here that CSR programs has been placed in the agenda items of a meeting to be attended by the senior management."*

Table 11.7 Barriers on the implementation of CSR practices

	Indicate whether you agree with the following statements	Agree	Neutral	Disagree
1.	Lack of management commitment to CSR	70 (80%)	10 (11%)	8 (9%)
2.	Lack of financial resources for CSR	78 (92%)	6 (9%)	2 (2%)
3.	Lack of knowledge about CSR	75 (85%)	9 (10%)	4 (5%)
4.	The complexity of CSR to implement	77 (88%)	8 (9%)	3 (3%)
5.	Lack of strategic plan for CSR	79 (90%)	7 (8%)	2 (2%)
6.	Lack of government support for CSR	77 (88%)	8 (9%)	3 (3%)
7.	Lack of regulations for CSR	74 (84%)	8 (9%)	6 (7%)

About strategic planning a respondent indicated, *"I can share my experience of multiple SMEs, which follow CSR without any clear strategic planning considerations. In the absence of such a framework, one can see it is very challenging for an SME to implement a CSR program successfully."*

On the complexity of CSR, one of the respondents concluded, *"The complexity of a CSR plan is also a thing that hinders our way towards its implementation. Being a small player, I find it difficult to get that level of complexity for effective implementation of a CSR plan."*

With reference to regulations, an owner stated, *"The regulatory framework for CSR deals only with the big companies. While SMEs appear to have an exception in this case. Being a small player, I do not remember anyone from any government agency approach[ing] us to ask such questions about CSR initiatives like sustainability."*

Conclusion

The objective of this paper was to identify and analyze the motivations, perspectives, practices, and barriers hindering SMEs from engaging in CSR activities in Nigeria. The results indicate that more than half of the respondents (60%) have heard of the concept of CSR. However, an overwhelming majority of the respondents does not understand the concept of CSR. In Nigeria, CSR activities are informal practices in the form of philanthropy and volunteer programs. The main motivations behind the adoption of CSR among Nigerian SMEs are increasing profits, attracting investors, enhancing community trust and support, favorable media coverage, pressure from stakeholders, and improvement of corporate image. Our results also reveal that the top barriers hindering SMEs from adopting CSR include a lack of management commitment, lack of financial resources, lack of knowledge, the complexity of CSR to implement, lack of strategic plan, lack of government support, and lack of regulations. These activities include charitable donations, community projects, volunteer activities, environmental projects, and being ethical. Respondents indicated that the perceived benefits of CSR practices by SMEs include improved organizational culture, attracting and maintaining the best employees, improved image and reputation, and improved customer loyalty in addition to reflecting the importance SMEs give to satisfying their most relevant stakeholders.

Implications

The results of this study have important theoretical and practical implications for contextual reasons. First, our results add to the small amount of research on this topic in Nigeria. By expanding studies beyond the developed Western world, the findings are particularly important because research of this nature is needed to further examine how and under what conditions CSR can be implemented in different contexts. In terms of policy implications, insights from this study can help the Nigerian government provide a more nuanced collaborative engagement with SMEs in addressing institutional gaps through CSR. In terms of practice relevance, we anticipate that this study will help SMEs address the barriers and challenges hindering them from engaging in CSR programs.

Limitations

This study has several limitations. First, the sample of the research was composed exclusively of selected SMEs from Nigeria. Therefore, the results of this study do not represent all SMEs in Nigeria and, as a result, cannot be generalized to countries or companies that are not part of this study. Second, given its structured survey design, our findings do not indicate what the responding SMEs are doing because the data is cross-sectional and does not cover overtime. Third, the SMEs surveyed and interviewed represent a small percentage, although moderately representative, of Nigerian SMEs. Hence, there is a need for future studies to explore the issues covered here in different parts of Africa to ascertain the extent to which the findings presented here are applicable to other parts of the continent. Similarly, future research can involve sectoral studies to determine whether SMEs operating within different sectors behave differently or similarly in their CSR practices.

References

Ahmad, N., Mahmood, A., Han, H., Ariza-Montes, A., Vega-Muñoz, A., Iqbal Khan, G., & Ullah, Z. (2021). Sustainability as a "new normal" for modern businesses: Are SMEs of Pakistan ready to adopt it? *Sustainability, 13*, 1944.

Arevalo, J. A., & Aravind, D. (2011). Corporate social responsibility practices in India: Approach, drivers, and barriers. *Corporate Governance, 11*(4), 399–414. https://doi.org/10.1108/14720701111159244

Brønn, P. S., & Vidaver-Cohen, D. (2009). Corporate motives for social initiative: Legitimacy, sustainability, or the bottom line? *Journal of Business Ethics, 87*(1), 91–109.

Business Day (2014). *SMEs and the CSR concept from* https://businessday.ng/brands-advertising/article/smes-and-the-csr-concept/

Bux, H., Zhang, Z., & Ahmad, N. (2020). Promoting sustainability through corporate social responsibility implementation in the manufacturing industry: An empirical analysis of barriers using the ISM-MICMAC approach. *Corporate Social Responsibility and Environmental Management, 27*, 1729–1748.

Carroll, A. B. (1991). The pyramid of corporate social responsibility: Toward the moral management of organizational stakeholders. *Business Horizons, 34*(4), 39–48.

Dahlsrud, A. (2008). How corporate social responsibility is defined: An analysis of 37 definitions. *Corporate Social Responsibility and Environmental Management, 15*(1), 1–13.

Fernando, J. (2022). *Corporate Social Responsibility (CSR) explained with examples*. Retrieved from https://www.investopedia.com/terms/c/corp-social-responsibility.asp

Freeman, E., & Moutchnik, A. (2013). Stakeholder management and CSR: Questions and answers. *uwf, 21*, 5–9. https://doi.org/10.1007/s00550-013-0266-3

Goyal, P., & Kumar, D. (2017). Modeling the CSR barriers in manufacturing industries. *Benchmarking International Journal, 24*, 1871–1890.

Graafland, J., & van de Ven, B. (2006). Strategic and moral motivation for corporate social responsibility. *Journal of Corporate Citizenship, 22*, 111–123.

Gupta, S., Nawaz, N., Alfalah, A. A., Naveed, R. T., Muneer, S., & Ahmad, N. (2021). The relationship of CSR communication on social media with consumer purchase intention and brand admiration. *Journal of Theoretical and Applied Electronic Commerce Research, 16*(5), 1217–1230. https://doi.org/10.3390/jtaer16050068

Hsu, J. L., & Cheng, M. C. (2012). What prompts small and medium enterprises to engage in corporate social responsibility? A study from Taiwan. *Corporate Social Responsibility and Environmental Management, 19*, 288–305.

Jansson, J., Nilsson, J., Modig, F., & Hed Vall, G. (2017). Commitment to sustainability in small and medium sized enterprises: The influence of strategic orientations and management values. *Business Strategy and the Environment, 26*(1), 69–83. https://doi.org/10.1002/bse.1901

Kumar, D., Goyal, P., & Kumar, V. (2019). Modeling and classification of enablers of CSR in Indian firms. *Journal of Modelling in Management, 14*(2), 456–475. https://doi.org/10.1108/JM2-07-2018-0100

Lam, J. S. L., & Lim, J. M. (2016). Incorporating corporate social responsibility in strategic planning: Case of ship-operating companies. *International Journal of Shipping and Transport Logistics, 8*, 273–293.

Lucky, E. O. I. (2018). An empirical examination of the influence of corporate social responsibility dimensions on the small and medium enterprise performance in Northern Malaysia. *International Journal of Business Innovation and Research, 16*(2), 243.

Lynes, J. K., & Andrachuk, M. (2008). Motivations for corporate social and environmental responsibility: A case study of Scandinavian airlines. *Journal of International Management, 14*(4), 377–390.

Mahmood, A., Naveed, R. T., Ahmad, N., Scholz, M., Khalique, M., & Adnan, M. (2021). Unleashing the barriers to CSR implementation in the SME sector of a developing economy: A thematic analysis approach. *Sustainability, 13*(22), 12710.

Merwe, M., & Wocke, A. (2007). An investigation into responsible tourism practices in the South African hotel industry. *South African Journal of Business Management, 38*, 1–15.

Meyer, M., Narjoud, S., & Granata, J. (2017). When collective action drives corporate social responsibility implementation in small and medium-sized enterprises: The case of a network of French winemaking cooperatives. *International Journal of Entrepreneurship and Small Business, 32*(1/2), 7.

Nejati, M., & Amran, A. (2009). Corporate social responsibility and SMEs: Exploratory study on motivations from a Malaysian perspective. *Business Strategy Series, 10*(5), 259–265.

OECD. (2005). ECD OECD SME and Entrepreneurship Outlook (3rd ed.). The Organization for Economic Cooperation and Development: Paris, France, p. 409.

Olusegun, A. I. (2012). Is small and medium enterprises (SMEs) an entrepreneurship? *International Journal of Academic Research in Business and Social Sciences, 2*, 487.

Perrini, F., Russo, A., & Tencati, A. (2007). CSR strategies of SMEs and large firms: Evidence from Italy. *Journal of Business Ethics, 74*(3), 285–300.

Pwc. (2022). *SME Survey in Nigeria*. Assessing current market conditions and business growth prospects. www.pwc.com/ng/en/events/nigeria-sme-survey.html

Reimer, M., Van Doorn, S., & Heyden, M. L. (2018). Unpacking functional experience complementarities in senior leaders' influences on CSR strategy: A CEO – Top management team approach. *Journal of Business Ethics, 151*, 977–995.

Ryan, R. M., & Deci, E. L. (2000). Intrinsic and extrinsic motivations: Classic definitions and new directions. *Contemporary Educational Psychology, 25*(1), 54–67.

Shen, L., Govindan, K., & Shankar, M. (2015). Evaluation of barriers of corporate social responsibility using an analytical hierarchy process under a fuzzy environment—A textile case. *Sustainability, 7*(3), 3493–3514.

Singh, K., & Misra, M. (2022). The evolving path of CSR: Toward business and society relationship. *Journal of Economic and Administrative Sciences, 38*(2), 304–332.

Sitnikov, C., & Bocean, C. (2017). Relationships between corporate social responsibility and strategic planning. In *Stages of corporate social responsibility* (pp. 121–137). Springer.

Tilley, F. (2000). Small firm environmental ethics: How deep do they go? *Business Ethics: A European Review, 9*(1), 31–41.

Yusliza, M. Y., Norazmi, N. A., Jabbour, C. J. C., Fernando, Y., Fawehinmi, O., & Seles, B. M. R. P. (2019). Top management commitment, corporate social responsibility and green human resource management: A Malaysian study. *Benchmarking: An International Journal, 26*(6), 2051–2078.

Zou, Z., Liu, Y., Ahmad, N., Sial, M. S., Badulescu, A., Zia-Ud-Din, M., & Badulescu, D. (2021). What Prompts Small and Medium Enterprises to Implement CSR? A Qualitative Insight from an Emerging Economy. *Sustainability, 13*(2), 952.

John O. Okpara, PhD, is a Professor of Management and International Business at the Zeigler College of Business Commonwealth, University of Pennsylvania, Bloomsburg, Pennsylvania, USA. He received his Ph.D. degree from New York University. Professor Okpara has held different academic and administrative positions in the higher education including the Chair of the Department of Management and International Business at the Zeigler College of Business for 10 years. He is a Visiting Professor of Strategic Management and the Coordinator of the Center for Management Training at the Warsaw University in Poland; Faculty Development Coordinator, Briarcliff College, Long Island, New York; Dean, Holy Trinity College and Theological Seminary, New York; a Visiting Research Fellow at the Nigerian Defence Academy (NDA), Kaduna, Nigeria; a Fellow of World Business Institute; and a Carnegie African Diaspora Fellow. He teaches courses in Global Business, International Management, Business Policies, and Strategic Management to undergraduate and graduate students. Professor Okpara's research interests include cross-cultural management, corporate governance, corporate social responsibility (CSR), strategic management of SMEs in developing countries, and social entrepreneurship and innovation. He has published several scholarly articles in several top management and international business journals, including Journal of World Business, Journal of Business Ethics, Management Decision, Thunderbird International Business Review, Corporate Governance, Journal of Management Development, Journal of Entrepreneurial Behaviour & Research, SAM Advanced Management Journal, Journal of Globalization and Small Business, among many others. He is also the founding Editor-in-Chief of the International Journal of Social Entrepreneurship and Innovation.

Dr. Jet Mboga is an Assistant Professor in the Management & International Department at Ziegler College of Business, Bloomsburg University of Pennsylvania. Her research interests include International Business: Sub-Saharan Africa; Entrepreneurship, Organizational Leadership Impact on Business and Society; Organizational Behaviour: Ethical Decision Making; Cross-Cultural Management; and Corporate Governance (Strategy). Her education is credited to PhD, University of Bolton, United Kingdom; Doctor of Business Administration Walden University, Minneapolis, MN.

Chapter 12
Responsible Entrepreneurship: Raising Corporate Social Responsibility Awareness Among Male and Female Owned and Managed Small and Medium-sized Enterprises (SME's) in Abeokuta, Ogun State, Nigeria

Adebimpe Adesua Lincoln, Brendhain Diamond, and Jane Croad

Societal Expectations on Corporate Social Responsibility

Various authors state that Corporate Social Responsibility (CSR) incorporates compliance with the law, philanthropy, and responsibility in the marketplace to various stakeholder groups who can affect or are affected by the achievement of the organization's objectives (Freeman et al., 2004). One of the earliest depictions of CSR was put forward by Howard Bowen (1953) in his seminal work in which he defines CSR as the obligations of businessmen to pursue policies, make decisions, or follow lines of actions which are desirable in terms of the objectives and values of society. Frederick (1960) opines that social responsibility concerns the ability of businessmen to oversee the operation of an economic system that fulfills public expectation and a willingness to see that society's economic and human resources are utilized for the wider society. According to Johnson (1971), a socially responsive entrepreneur or manager is interested not only in his own well-being but also that of the other members of the enterprise and his fellow citizens. Jones (1980) views CSR as a voluntary obligation owed by an organization to constituent groups within society

A. A. Lincoln (✉)
The School of Law and Social Justice Building, University of Liverpool, Liverpool, UK
e-mail: laulinc@liverpool.ac.uk

B. Diamond
University of Liverpool, Liverpool, UK
e-mail: b.diamond@liverpool.ac.uk

J. Croad
Robert Kennedy College, Zurich, Switzerland
e-mail: jane.croad@rkc.edu

U. Nwagbara et al. (eds.), *Corporate Social Responsibility Disclosure in Developing and Emerging Economies*, CSR, Sustainability, Ethics & Governance, https://doi.org/10.1007/978-3-031-61976-2_12

other than shareholders such as customers, employees, suppliers, and the wider community.

Societal views are a critical factor within many CSR perspectives. Wood (1991) and Carroll (1979) discuss CSR from within a context of society and its relationship with companies. Both authors argue for companies to proactively perform CSR actions based upon the assumption that society expects companies to do so for the benefit of various stakeholder groups within society. Thus, examining the nature of what society expects would be a logical next step in determining whether a consensus exists within society to identify a society's views. For Shaw (1988), the stakeholder theory will maintain its dominance as long as society sees the government as capable and successful at dealing with social ills. Therefore, society recognizes that someone is responsible for caring for society's needs, and, primarily, this responsibility is placed upon governments. Only when governments fail in this pursuit, for example, the failure of successive Nigerian governments where CSR and societal development/advancement is concerned, would citizens then place this burden upon others, such as companies. In other words, as long as governments continue to service the population's needs, there is no need for a Kuhnian paradigm shift. Kuhn posits that where corporate paradigm of the single profit aim is not functioning satisfactorily a replacement is needed (Kuhn, 2012).

Shaw (1988) sees Freidman's shareholder primacy perspective as hinging upon society's views on Western governments' relative success in dealing with society's well-being. The reasoning was that countries where the governments effectively look after society's well-being will see less need to place part of the responsibilities upon companies. At the very least, this implies that companies would not have an imperative of responsibility unless the government cannot fulfill its role. The result is that different societies and groups within society may have different views and expectations on companies and their responsibility toward the general well-being within society, as expectations have a subjective nature. By reviewing different groups within society, an understanding of the general expectations may be deciphered.

Theoretical Perspectives

The review in this section adopts a varied theoretical approach. Various theoretical perspectives such as Carroll's pyramid, instrumental theory, stakeholder theory, legitimacy theory, ethical theory were considered in providing a holistic theoretical underpinning.

Carroll's Pyramid

CSR is often considered as a firm's obligation for its economic, legal, ethical, and philanthropic activities (Carroll, 1991). This view was echoed by McGuire (1963) when he asserts "corporations have economic, legal obligation and responsibilities to society at large which extends beyond these obligations" (p. 144). Carroll recognizes that companies were formed to make a profit while efficiently supplying society with its needed goods (1991), thus the pursuit of profit (economic) is the foundation upon which the pyramid of responsibility is based and represents the first level of Carroll's CSR pyramid. The second level in Carroll's CSR Pyramid represents laws. Laws codify what society sees as crucial moral rights and wrongs. Laws being in place can be likened to Holmes's do not be a "bad man" (Holmes, 1897). The third level of the pyramid extends company responsibility to following ethical norms that are not codified in law. Therefore, even when the law does not dictate it, corporations ought to act fairly and do what society sees as right and just (2004).

For Carroll, corporations need to be moral creatures and act conductively to broader societal norms. Thus, managers should not be amoral by only making decisions while adhering to the legal requirements. Instead, managers should decide based upon what is morally just. At the very least, this would include "doing no harm," but being fair also could include not abusing others when an imbalance of power exists. Carroll's fourth level of CSR (i.e., philanthropic) is not mandatory but sees companies as having the option to choose to be "good men" and act as a good citizen who act within a philanthropic sense. This level could translate into companies contributing to the well-being of stakeholders in the broad sense of the term. Carroll recognizes that with multiple responsibilities comes conflicts, as managers must make complex decisions while considering the four levels and seek a mutually beneficial solution. Carroll does acknowledge that changing situations would affect which responsibilities would take precedence in decision-making. It follows from Carroll's views that, in general, managers would need to act not only as amoral decision-makers based upon economic criteria but also become moral creatures.

Instrumental Theory

Instrumental theory limits responsibilities upon the traditional and narrow view of the function of companies as focusing upon seeking profits for company owners. As they describe, the instrumental approach does not exclude companies from "doing good" toward society as long as doing so adds to the owners' wealth and does not detract from profit maximization (Garriga & Melè's, 2004). Milton Friedman can be seen as advocating this approach and focuses on management's agency responsibilities to owners (Friedman, 2002). Instrumental theory can be described as mono-dimensional in a company's relationship with society in that it is limited to the economic dimension. All societal interactions are to be seen within the context of the pursuit of shareholders' wealth. Fry et al. (1982) compared tax records from 1946 to 1976, including 36 industries, and identified a positive correlation between advertising expenses and corporate philanthropy during the 30 years. These findings would seem to indicate the use of CSR in pursuit of increased profits. Additionally, companies in more contact with the public spent more funds for philanthropic endeavors. These findings led the researchers to conclude that companies' philanthropic activities were profit-motivated and within the interests of the shareholders. Therefore, managers are not acting out of an enlightened sense of responsibility to the greater society, but they are more enlightened about other ways to increase company profits beyond cost savings and traditional promotional advertising. It is noteworthy that this study was conducted over 40-years-ago before many of the CSR theories had come to prominence. That said, the study gives evidence that companies were historically operating in the instrumental theory. It might be helpful to perform this same analysis on contemporary corporations to see if this still holds.

Political Theory

Political theory recognizes that companies have de facto power within society due to their wealth and influence. Advocates of political theory argue that with power comes the responsibility to be proactive within social and political dynamics (Garriga & Melè's, 2004). Therefore, companies' great power must be wielded responsibly and for the benefit of others. Davis (1960) saw two types of duties that managers had beyond the pursuit of profit: socio-economic—an economic benefit to society, and socio-human—contributing to the moral development of society. While companies have power in society, Davis recognizes that managers wield influence through their decisions. Davis did see the pursuit of profit as still the primary aim for companies, but the need to act with a greater responsibility "do(es) exist." Davis warned that if companies do not take responsibility themselves, their social power would be lost, whether taken by competitors or the government. Davis assumes that societal norms place expectations upon companies, or at least society sees CSR positively enough to influence consumers to act upon this favorable view when they purchase products. This assumption may well not be the case, as will be discussed later in this work. If these assumptions are valid, then it is in the interest of companies to take greater responsibility. Davis saw that the first social duty of companies is to figure out a plan as to how they can act with social responsibility, despite managers lacking the skills and knowledge to wield social power responsibly. Furthermore, managers are not likely to be well suited to decide what moral development other members of society should have. Companies could open themselves up to backlashes from consumers who disagree with companies' socio-human actions to change societies' morals.

An additional and prominent concept within political theory is corporate citizenship (Garriga & Melè's, 2004). Implied within the term, corporate citizenship is the recognition that companies should be seen similarly as ordinary citizens within society. Matten et al. (2003) provide an overview of the concept of corporate citizenship in which there are three main types: a narrow view, a broad view, and a CSR equivalent type (Matten et al., 2003). With the CSR equivalent type, there is a conflation of corporate citizenship with definitions of CSR. Matten et al. see the narrow view as companies having the option to take philanthropic actions in much the same way as any other citizen. While there is no mandate for them to act in society's interest when making business decisions, management may choose philanthropic activities using shareholders' funds to benefit the owners. Therefore, social spending can be within the enlightened self-interest of the company owners. Moir and Taffler (2004) argue that this is not true altruism but a strategic move within self-interest, thus hardly manifesting in the company being an ideal citizen. This same tactic is used effectively by billionaire criminals and criminal gangs to counterbalance their objectionable (i.e., "bad man") activities (Kuldova, 2018). Matten et al.'s broader view of corporate citizenship hinges on the rise of neoliberalism and the associated reduction of government activity in looking after the population's social needs, combined with globalization's effect of reducing the influence of government authority. With governments becoming less significant, someone or thing must fill their role. As companies are powerful, then they are seen as being responsible for acting in a socially responsible role. In developed countries, Matten et al. see corporate citizenship as an attempt to transfer this role to companies via a self-interest motivation. On the other hand, they recognize that developing nation governments lack the resources to take on the full responsibility of providing these benefits to their citizens. Therefore, under the broader type, companies would take responsibility for the activities that governments are incapable of performing.

Integrating Social Demands Theory and Stakeholder Theory

Integrating social demands theory centers around companies' responsibilities stemming from them being dependent upon society for their existence and operation. As such, the company's legitimacy depends on its responsiveness to the ever-changing dynamic within society and society's expectations of the company (Garriga & Melè's, 2004). Stakeholder theory could fall into this theory type, as stakeholder theory is based upon companies recognizing that they are not an island unto themselves. Instead, companies operate within society, and company activities affect others in society. When these impacts are negative, then society may see the company similar to how they see a "bad man." As Holmes pointed out, when society does not like specific actions, it creates laws to restrict these actions. In other words, actions that society sees as not legitimate become illegal. Stakeholder theory and integrating social demand theory both recognize the inherent responsibility upon companies to act with consideration to all those affected by the company. This recognition can be seen as an acknowledgment of the interdependency between companies and the population. In effect, companies must do as society wants them to do because society enables and permits the company to exist. Should the company act outside of what society sees as legitimate, then society may bring about laws restricting the company or end the company's existence. A similarity can be drawn between this view and a child being under the authority of a parent. Parents may expect children to behave in a way that the parent approves. "I made you, I feed you, and you live in the home that I pay for; therefore, you need to do what I expect of you!" In effect, this approach is discounting the owners' role, purpose, and interest in the company.

Ethical Theory

Ethical theory has a philosophical base that sees companies as operating within societies governed by ethical norms. Therefore, companies should also follow society's ethical rules. The authors present a variety of ethical theory approaches from "do no harm," "net positive," to "only positive impacts" for the common good of society (Garriga & Melé, 2004). Ethical approaches can be seen as striving to create a moral philosophy as the foundation for the approaches discussed above. With morality as its base, ethical theory would need to reflect the moral expectations that members place upon each other and, specifically, on companies. Therefore, this theory would be limited by the subjective nature of morality, and the validity of a theory would be dependent upon the subjective and changing ethical norms in a society. The subjective nature of ethical approaches could be used to validate actions reflecting the ethical norms in the different jurisdictions and societal context in which the firm operate. If corruption is routine in one society, the firm could adapt these norms and feel justified in participating in the corruption, as they are simply following local ethics.

Limited Views and Lexicographic Utility Theory

In addition to the above approaches outlined above, some academics hold much more limiting views. Milton Freidman's view, wherein companies are only responsible to the owners, could also be included in this group. Freidman's view would then place no social responsibility on the company but sees the company as contributing to the well-being of society via jobs, products, efficiencies, and paid taxes. Direct social spending would then be counterproductive and a disservice to the owners. Harold Johnson offers a more moderate view by extending Freidman's limitation of primacy to owners and also allows for companies to take "into account employees, suppliers, dealers, local communities, and the nation" (Johnson, 1971). Johnson did see companies as being open to focusing upon maximizing company profits and including other societal goals. Johnson saw philanthropic actions as being justified when directly connected with increasing profits for the owners, as seen in instrumental theory. The Lexicographic utility theory was described by

Johnson as applying when companies had achieved their internal profit goals, and then the company takes on behavior as if they have extended social responsibilities, despite them not actually having responsibility. For example, a company that has achieved beyond its profit goals and then donates some surplus profit to a society-servicing charity. This charitable donation might be misinterpreted by some as the company recognizing social responsibility when it does not see itself as being responsible. The company is doing good without feeling the burden to do good all of the time.

What emerges from this discussion is a lack of consensus in almost all areas. The academic literature is divided upon what should motivate companies: altruism, self-interest, or moral obligation. There is also division upon the extent to which companies can be expected to act, ranging from only in the owners' interests to only having positive impacts as a good community member. As such, it cannot be definitively stated that the academic literature expects companies to act with social responsibility beyond the minimum requirements of the law or even when it could be within the interests of the shareholders seeking profits. In general, these theories and promoters of these theories do not derive their conclusions from observing societal norms and then forming a model of these norms. Instead, the promoters tend to present their theories as ideals of how they see societies and companies "should" be and "should" act with little or no regard for society's pragmatic dynamic norms. While there are some exceptions, such as Fry et al. (1982), most authors' views are not validated via existing norms but present a theory about how CSR responsibilities can be justifiably placed upon companies. As their theories depend on societal norms, their end conclusions can be seen as dubious unless societal norms place CSR expectations upon companies.

Morality and Societal Norms as Reflected in Law: Stakeholder and Legitimacy Theory

As the above discussion highlighted, the validity of these conflicting theories hinges upon the norms within society as to its subjective expectations upon companies. In order to satisfy society's expectations, one must understand what society expects or has as its ethical norms. For Hart, in his book The Concept of Law (2012), social rules only exist when at least some people in society expect adherence. Therefore, social rules can exist in subgroups in society even when they are not generally accepted. Stakeholder theory claims to be grounded upon general society norms, while the expectations within legitimacy theory need only be held by the company and not necessarily by the general society. Hagerström and Olivecrona (1953) see social norms as the expressions of the will of society as a whole and carrying a general expectation for actors to conform, and non-conformance would be generally considered "bad behavior." Thus, for a social norm to exist, the expected behavior must be supported by, as Hart (2012) puts it, "a distinctive attitude to that conduct as a standard." Holmes (1897) saw the law as an extension of society's moral norms, as laws are created with the intent of maintaining order within society. For Holmes, laws are not aimed at "good men" as "good men" already conform to societal norms.

Conversely, laws are aimed toward "bad men" whose behavior may otherwise be unconducive with societal norms. On the intent of laws, Holmes' views do not directly conflict with shareholder, stakeholder, or legitimacy theories as each of these theories allows, to some degree, for laws to reflect societal norms, with stakeholder and legitimacy theory emphasizing what companies ought to do. Davies (1973) agrees with Holms' view and extends the view to include that every good citizen does the minimum but sees companies' social responsibility as necessarily going beyond the legal minimum requirement. He predicates this obligation as being based upon companies having disproportionate power and influence within society (Davis, 1960). Thus, Davies argues that companies should take proportionate responsibility, they must be more than a "good" citizen for a company to be socially responsible. Arguably, Davies' position would place the burden upon companies to be "ideal" citizens and not simply good citizens. A more extreme form of the ideal citizen is in Pristine Business Goals, wherein profits are the secondary aim behind fulfilling all societal groups' needs and ensuring their basic needs are satisfied (Hoque et al., 2018). Davies' approach is conducive with both stakeholder theory and legitimacy theory as these are not limited by applicable laws but place responsibility for good behavior upon companies. Arguably, Davies' view is more compatible with legitimacy theory in that this theory prescribes companies to proactively take responsibility beyond what is expected of citizens in general.

The positions of the above three authors lend insight into the relationship between societal norms and laws as reflections of these norms. It is noteworthy that Hagerström differentiates between legal rules and social norms as not necessarily being the same or being perceived in the same way. This mainly hinges upon feelings of obligation and the motivations to comply wherein laws carry stronger rational obligations are motivations themselves, while social norms can provoke varying degrees of subjective feels of obligations and motivations. When considering the fluctuating nature of societal expectations, it may be reasonable to anticipate a lag between societal norms and enacting laws reflecting these norms. This lag time could be an essential consideration for both stakeholder and legitimacy theories in that they prescribe companies go beyond the minimum legal requirements.

Laws can be seen as coming in two primary forms: positive laws and negative laws (Schuster & Rachlin, 1968). Positive morality in making persons proactively "do" something, as opposed to the negative of preventing them from doing something. Within this framework, positive laws prescribe what actors are "to do," while negative laws prescribe what actors are "to not do." "To do" something would mean that a law requires actors to perform a particular action; for example, to use a seatbelt while driving. On the other hand, "to not do" could be a requirement to not drive faster than the speed limit, which can be written as a positive law stating that one is to adhere to the speed limit. Laws binding fiduciary duties of company directors to the shareholders can be seen as both requirements to do and not do. Dodge v Ford Motor Co (1919) recognized the director's duties to act for the shareholders' benefit and not take actions contrary to shareholder's profits interests. This ruling supports Freidman's view of companies' responsibilities being limited to profit for shareholders, as directors should not do anything to detract from this pursuit.

On the other hand, morality could have a third element wherein society may expect actors "to not not do," in other words, what one "ought to do" (Keke, 1984). This moral "ought" carries an implied voluntary element as there is no legal requirement to act, but failure to act might bring about judgment from a society that the non-acting individual's inaction was "bad behavior." Meaning, others in society may hold the expectation to act despite there not being a law that one must act. For example, in the case of someone drowning, there is a rule compelling individuals to lend assistance, the failure to assist could attract scorn from society. Keke (1984) recognizes that the obligation of "ought" should be conditioned by the ability to act or "can." Therefore, the expectation of "ought" depends on the capacity of the individual to have the capacity to have acted. For example, a toddler would not be burdened with "ought" in the case of someone else drowning because the toddler would lack the capacity to understand the situation and to lend assistance. At the same time, an adult who lacks the capacity to save the drowning person, perhaps they themselves cannot swim, then mediates the burden of "ought." This third form within morality lends itself to the legitimacy theory in that since it is assumed to be within the power of large companies to aid others in society, to not do so would be viewed as Hagerström's bad behavior. Large and financially wealthy companies when compared to their micro and small counterparts would be expected "to not not do" something and, therefore, if they were able, then they "ought" to aid others in need within society. If such companies do not act, then they could be seen as having reduced legitimacy to operate. Indeed, many academics see companies as being obliged to act by aiding those who are less fortunate, as to Not act would be considered "bad" (Weyzig, 2009). These actions do not necessarily result in the company "being good," as companies doing good philanthropic acts can also be found doing bad actions in the pursuit of profit (Norman & MacDonald, 2004). Additionally, doing what everyone "ought" to do is expected and could be viewed as a minimum requirement within society. Mandating corporations to lend help only because they represent the concentrated wealth of the owners is distinctly different from mandating a medical doctor to lend help. Good Samaritan laws aim to reduce legal liability by protecting those who volunteer help from being prosecuted, while these laws predominantly only compel people with specific medical certifications to do what they "ought" to do (McQuoid-Mason, 2016).

The foregone discussion would indicate that consumers and society are fickle when it comes to CSR. The discussion also brings into question the conclusions of surveys concerning society's "expectations" upon companies. Since consumer buying behavior is not aligned with the survey results, the problem is possibly in the surveys themselves. While consumers may be "favorable" to the general idea of companies performing CSR activities, consumer behavior seems to indicate that this favorable view does not reach the level of "expectations," as consumers are not generally influenced by positive CSR actions taken by companies "to do good." Consumers may like the idea, but consumer spending patterns show that they do not expect positive CSR actions from companies. Thus, consumer behavior does not prompt a paradigm shift within company culture because as long as companies are not doing bad, there is little need for them to do good. The very subjective nature of individual consumers' views invites variation of expectations between different groups within society. For example, the Millennial Generation is generally more engaged in social justice actions with higher expectations placed upon companies to behave with higher ethics (Andersson, 2018). Some of this may qualify as expectations as Millennials are more likely to invest in companies with positive ethics profiles and avoid unethical companies (Logsdon & Van Buren, 2008), but this does not mean that all members of this generation invest based upon CSR expectations, which reflects a lack of consensus within the group.

Overall, the discussion does support the view that companies are "to do no harm" as consumers expect this, else the company may suffer a loss in sales and associated decreased profits. Laws requiring companies to act responsibly by not damaging the environment or violating human rights would seem to match societal expectations, while corporate philanthropy should not be seen as being expected. This dynamic can undermine the view of extending CSR where companies are responsible for solving general societal welfare. There is little reason to believe that, in general and collectively, individuals in society expect companies to pursue philanthropic CSR proactively. Society may be favorable to the idea in surveys, but they infrequently demonstrate this in their action or spending. Additionally, although some in society can expect companies to do what they ought to do toward society's general well-being, this does not reach the level of being a general norm as there continues to be significant variations within society.

Corporate Social Responsibility and Responsible Entrepreneurship

SMEs are an important engine of growth in both developed and developing countries. These firms help alleviate poverty and are key contributors to job creation, innovation, and economic growth and development (Woldie et al., 2008; Adesua-Lincoln, 2011). This study therefore seeks to raise awareness of CSR among SME's in Abeokuta and assist the SME's in thinking about their firm's efforts toward responsible entrepreneurship by raising questions on the possible ways SME's could improve their business in a profitable and sensible manner. The study also seeks to assist the SME entrepreneurs identify further actions to take to strengthen their business reputation and performance. SME entrenchment in the local socio-economic environment, their interaction with clients and employees in the local communities make them key contenders where CSR is concerned since a good reputation is essential to achieving competitive advantage. SME's are also more aware of local practices, associated risks and emerging social issues within their local communities than large corporations. Consequently, SME's are more likely to conduct their business in a socially responsible manner, respond to social and community needs, and engage in activities, which take into consideration key stakeholder interests. CSR is important to all firms large and small, and businesses have the responsibility of aligning diverging interests of stakeholders by effectively adopting appropriate strategies. SME approach to CSR differs significantly from the approach adopted by larger firms (Lincoln, 2017). This is due in part to the nature of the SME firm and associated characteristics such as size, sector, and the SME owner/managers' orientation toward profit (Braun, 2010). SME firms' ownership and management are not separated as such control remains in the hands of the owner (Sweeney, 2007). It stands to reason that SME's acceptance of CSR is largely a factor of the personal attitudes of the owner/manager. Furthermore, due to the informal hierarchical structure of the firm, SME's are more suited to fostering effective networking, interpersonal relationships, and cooperation based on trust of stakeholders and legitimacy. Furthermore, the SME owner/managers' personal values, ethical as well as moral convictions to an extent impact CSR strategies adopted in light of the lack of separation between ownership and control of the business (Lincoln et al., 2015; Lincoln, 2017, 2018).

In spite of the associated benefits, which can be reaped from SME engagement in CSR related activities, evidence shows that many SME's are unaware of CSR initiatives (Jenkins, 2006; Lincoln, 2017). A review of the literature highlights that CSR and social community activities undertaken by SME's tend to be fragmented and informal in nature (Jamali, 2009; Maitland, 2002; Perrini et al. 2007; Lincoln, 2017). Lepoutre and Heene (2006) suggest that SME's seldom engage in formal CSR strategies. SME's are also said to encounter greater barriers and challenges in their quest to engage in CSR activities and are prevented from active CSR engagement due to various factors ranging from their size, lack of time and support, lack of financial resources and expertise on CSR issues (Blomback & Wigren, 2009; Lepoutre & Heene, 2006; Ortiz & Kuhne, 2008; Lincoln et al., 2015; Lincoln 2017). For example, Hsu and Cheng (2012) state that many SME's consider that they lack the requisite skills and guidance in CSR, this is because some of the guidance developed on a global level or sector specific level are not suitable for their purpose. Murillo and Lozano (2006) state that SME's lack knowledge and know-how in relation to CSR issues which act as a barrier to their ability to effectively engage in CSR. Furthermore, SME's make little or no attempt to communicate their involvement in CSR related activities or publish annual reports, which discloses their CSR engagements (Fassin, 2008; Giovanna et al., 2012; Murillo & Lozano, 2006). Fassin (2008), however, assert that SME's actively engage in socially responsible practices although they do not formally publish or report their activities.

Corporate Social Responsibility Regulations and Initiatives in Nigeria

Despite all of the CSR activities, there has been no real change in the approaches and attitudes within the companies concerning environmental and social impacts. The CSR actions allow for the impression that the companies are taking responsibility, while a lack of internal monitoring means that these companies take on no broader responsibilities. Comparatively weak governments in developing nations either lack CSR rules or the ability to enforce them and often lack both (Dine, 2012). In Nigeria's case, the social instability in the region may also make stronger regulations and accountability unsustainable. There is the additional fear that if governments place high regulatory burdens upon businesses, concerning social responsibilities and accountability, then the additional burden may stifle firm growth and survival. It is imperative that any CSR framework adopted is intended to promote SME advancement. There are arguments for voluntary and mandatory CSR compliance. Voluntary compliance affords businesses a level of control as to which CSR initiatives to adopt. These initiatives can be tailored by the SMEs to suit the nature of their business activities, their customer base and ensure that the SME's adopt practices that provide maximum benefits to the local communities in which they operate. Mandatory compliance, on the other hand, requires legislative intervention and control in ensuring transparency and accountability. While this approach may ensure that more businesses are CSR compliant and afford a holistic and uniformed approach to CSR practices adopted by businesses across all sectors and industries, there are cost implications not only to the businesses but also to the government parastatal responsible for its enforcement. In addition, uniform enforcement may prove burdensome to an already fragile legislative landscape. There are also issues vis a vis the complexity inherent in drafting the core tenants of an all-encompassing CSR framework.

In 2007, Nigerian lawmakers attempted to pass the Corporate Social Responsibilities Bill. The Bill contained mandatory requirements aimed at community development within the communities in which companies operated to compensate these communities for the negative consequences of company operations. A feature of the Bill was that the CSR activities were to be regulated by a government CSR Commission, much as the UK formed in 2002 and later disbanded in 2010. Dine (2012) cataloged some of the bill's shortcomings as it being seen as placing control in the hands of government officials with little knowledge of the company operations, the inefficacy of the Nigerian government to regulate corporations, and it was de facto an additional corporate tax of 3.5%. In effect, the Bill was seen as oppressive, intrusive, and untenable in practical application. In 2020, Nigeria updated its primary corporate legislation Companies and Allied Matters Act (CAMA). Sect. 305 (3) of CAMA states: "directors shall act at all times in what he believes to be the best interest of the company as a whole …and in doing so, shall have regard to the impact of the company's operations on the environment in the community where it carries on business operations." In addition, Section 305 (4) of

CAMA states that "The matters to which a director of a company is to have regard in the performance of his functions include the interest of the company's employees in general, as well as the interest of its members." This is a welcome development in enshrining the enlightened shareholder perspective into the fabric of corporate law. That said, it appears that the provision embraces facets of the shareholder centric notion of wealth maximization as such limited in its application in light of the provisions contained under Section 305 (9) of CAMA which states that "any duty imposed on a director under this section is enforceable against a director by the company." The exclusion of other key players in exercising the rights under Section 305 (3) and (4) is a major weakness of the 2020 Act. It is worth nothing that Nigeria does not have a codified law regulating CSR. Furthermore, no definition exists on the concept. Consequently, it is unclear what amounts to CSR within the Nigerian context. Anecdotal evidence suggests that CSR in the Nigerian context incorporates socio-cultural, ethical, and moral norms.

Research Methodology and Regional Context

Data for the study was collected from 72 SME's in the Abeokuta region of Ogun State, Nigeria. Ogun State was selected as the apt location for the study in light of its proximity to Lagos State which is the commercial nerve center of the country and recent developments carried out in the city; for example, the network rail service linking the city with Lagos and Ibadan. Abeokuta is the capital of Ogun State and the largest urban center, it is located in the South West Geopolitical Zone of Nigeria. Ogun State lies to the north of Lagos State, to the west of Ondo State, and to the south of Oyo State and Osun State (UN, 2020). Abeokuta is built in the center of the Lagos-Ibadan extended urban region and forms part of the larger metropolitan economic area. This strategic location along with the presence of diverse local resources, rapid population growth, and enhanced political status has generated dynamic economic activities (UN, 2020). A few large-scale industrial establishments, as well as trade, personal services, finance and insurance services comprise the local economy. Abeokuta is also an agricultural trade center and an exporting point for various commodities. The population of Abeokuta is growing rapidly, mainly due to natural causes, immigration, and rural-to-urban migration (UN, 2020). The population almost doubled from about 375,000 in 1991 (according to the population census) to over 700,000 in 2013. Due to rural-urban economic exchanges and part of the population residing outside the urban area, day and night population shifts require commuting between urban and rural areas of the state (UN, 2020).

Abeokuta is witnessing rapid urbanization and real estate expansion in recent times due to heavy investment in infrastructure development projects by the Ogun State and Federal Governments. Investments in road, rail, and aviation transportations have resulted in innovation and gentrification of parts of Abeokuta. Abeokuta was identified by United Nations Habitat (UN Habitat, 2018) as a core location to contribute to sustainable urban development based on the 2030 sustainable development goals to contribute to the mobilization effort to end poverty, fight inequalities, and tackle climate change. The identified projects for Abeokuta are: (i) an Urban Master Plan and Guidelines for Urban Renewal and (ii) a Public Transport Policy. Abeokuta is targeted for sustainable urban development in the following key SDG areas:

- Goal 5 and 8 inclusive economic growth through gender equality and decent work and economic growth. The focus here also includes the promotion of inclusive economic growth and employment creation, integration of the informal economies and poverty reduction. In particular, there is also a focus on enhancing access to economic resources for women and vulnerable groups.
- Goal 6 and 7 basic service for all through affordable and clean energy and clean water sanitization. Of importance is the need for the introduction of sanitation, clean water, affordable, reliable, and sustainable energy for all and the protection and restoration of water ecosystems.
- Goal 9 and 10 reduced inequalities and industry innovation and infrastructure. The focus here is on the implementation of quality, reliable, and sustainable and resilient infrastructure which supports economic development and well-being of the people of Abeokuta.
- Goal 11 sustainable cities and communities. The focus here contributes to the promotion of safe, adequate, and affordable housing, sustainable transport, and safeguarding cultural and natural heritage including access to green and public spaces.
- Goal 17 effective institutions through partnership for the goals. This also focuses on capacity building of local institutions and knowledge sharing (UN Habitat, 2018).

Questionnaire surveys were used to collect data for the study. The questionnaire sought to measure the SME's efforts toward responsible entrepreneurship and identify further actions entrepreneurs could adopt in strengthening their business practices, reputation, and future performance. The entrepreneurs who took part in the study were identified using convenience sampling. The sample includes male and female owners of firms in different industries, different trading companies, and companies in different geographical locations throughout Abeokuta in order to ensure diversity. Data obtained from the questionnaire was analyzed using JASP Open Access Statistical Analysis Package for Mac. Descriptive and inferential statistical methods were employed. Exploratory factor analysis was used as a data reduction tool to identify the most important CSR and responsible entrepreneurship issues. Percentage value was used in analyzing and presenting the initial results vis a vis the demographic characteristics, CSR, and responsible entrepreneurship. Exploratory factor analysis was used as a data reduction tool to identify the most important CSR and responsible entrepreneurship factors. The researchers at all times ensured the integrity of both secondary and primary data. Data collection was conducted in a manner which ensured ethical issues were taken into consideration, and that the research participants were protected at all stages of the research process

(Saunders et al., 2015). Informed consent was obtained from each entrepreneur who confirmed their understanding of the nature of the research and their voluntary participation in the research.

Analysis and Discussion of Findings

The result is presented in two parts. Part one focuses on the entrepreneur's awareness of CSR and responsible entrepreneurship vis a vis policies and practices in five main aspects, namely, workplace and employees; environment and climate; marketplace; local community; and company value. In part two, the result of the exploratory factor analysis is presented. Exploratory factor analysis is used as a reduction mechanism in relation to the five main aspects identified above. The reduced number of factors obtained from the analysis and the underlying dimensions between measured variables and latent constructs enables the formation and refinement of theoretical perspective on divergent perceptions of the SME's approach to CSR and responsible entrepreneurship. The Cronbach's Alpha test was used to statistically test the internal consistency of various parts of the questionnaire (Bryman & Bell, 2018). The Cronbach Alpha coefficient in the questionnaire is 0.92. As the internal reliability value is above 0.80, the study has a satisfactory level of internal consistency as shown in Table 12.1 below.

CSR and Responsible Entrepreneurship in Abeokuta

In total, 37 male entrepreneurs (51 percent) and 35 female (49 percent) entrepreneurs took part in the study. The questions sought to evaluate the entrepreneurs' CSR and responsible entrepreneurship policies and practices, identify areas in which awareness of CSR and responsible entrepreneurship is required.

Workplace and Employee Policies and Practices

The questions in relation to workplace policies and practices sought to raise awareness of various employment relations issues such as employee training/performance appraisal, discrimination in the work place, health and safety, employee

Table 12.1 Cronbach's Alpha and Gutmann's $\lambda6$ reliability statistics

Cronbach's Alpha	Gutmann's $\lambda6$	N of Items
0.92	0.992	72

Table 12.2 Workplace and employee policies and practices

CSR and Responsible Entrepreneurship: Workplace Policies and Practices (WPP1–5)	Male Percentage	Female Percentage	Cumulative Percentage
WPP1: Developing skills and long-term careers goals by encouraging employees through training plan/performance appraisal	84%	91%	86%
WPP2: Workplace policies to ensure adequate steps are taken against discrimination in recruiting staff and in the workplace	81%	86%	83%
WPP3: Employee consultation on important issues	68%	80%	74%
WPP4: Suitable health, safety, and welfare plan for employees	65%	74%	69%
WPP5: Work–life balance for employees through flexible working hours or working from home	62%	69%	65%

consultation, and work–life balance. The findings show that the majority of the entrepreneurs were aware of the importance of employee's well-being as a strategic tool in the long-term success of the business venture. The finding in Table 12.2 below shows a higher representation of female entrepreneurs stated that they consult their employees on important issues, adopt suitable health, safety, and welfare plans, and provide employees with flexible working hours compared to male entrepreneurs. Employee protection and employment law provisions in Nigeria lack vigor when compared to some of the provisions in countries such as the UK or other European Countries. In many cases, employees often work in environments that are hazardous to their health and safety and in many cases endure late payment of salaries. Many businesses large and small are often dependent on the knowledge and skill as well as the talent and creativity of the employees. Providing employees with basic needs, clear path to career progression, training and development opportunities as well as actively allowing employees to contribute to decision-making ensures that the workforce is committed, empowered, and loyal. This in turn enhances organizational efficacy. In addition, a close nexus is required between workplace and employees and the company value. It is imperative that entrepreneurs introduce their workforce to the core values of the organization and that they offer ongoing training and induction programs to ensure employees are familiar with the organization's values and ways of working.

Environment and Climate Policies and Practices

All business practices have implications for the environment and our climate. Environmental degradation and the climate crisis are now at the forefront of the global agenda. Consequently, businesses large and small are encouraged to design and implement robust environment and climate policy and practice which focuses

Table 12.3 Environmental and climate policies and practices

CSR and Responsible Entrepreneurship: Environmental and Climate Policies and Practices (EP6–14)	Male Percentage	Female Percentage	Cumulative Percentage
EP6: Reduces environmental impact through energy conservation	49%	69%	58%
EP7: Reduces environmental impact by engaging in waste minimization and recycling	68%	77%	72%
EP8: Reduces environmental impact by pollution prevention through emissions to air and water, effluent discharges, noise	68%	86%	76%
EP9: Reduces environmental impact by protecting the natural environment	86%	94%	90%
EP10: Reduces environmental impact by utilizing sustainable transport options	49%	63%	56%
EP11: Engages in active assessment on environmental impact through recycling, reducing energy consumption, preventing pollution	65%	77%	71%
EP12: Considers potential environmental impacts when developing new products and services; for example, assessing energy usage, recyclability, or pollution generation	54%	71%	63%
EP13: Supplies clear/accurate environmental information on products, services, and activities to customers, suppliers, and local community	57%	71%	64%
EP14: Seeks to adopt practices where the enterprise could use the sustainability of its products and services to gain an advantage over competitors; for example, recyclability of products, energy efficiency	68%	80%	74%

on reduced energy consumption, reduced pollution, and adoption of efficient recycling systems and protection of natural habitats. The findings in Table 12.3 below show that the entrepreneurs were aware of current global drives where climate change and environmental conservation are concerned and were actively engaging in ensuring that their business practices have negligible environmental and climate impact. Significant variations can be seen in the responses obtained from the male and female entrepreneurs. More awareness and effort is required among male entrepreneurs in relation to efforts to reduce environmental impact through energy onservation and sustainable transport options that can contribute positively to organization success. There are financial benefits which can be gained from adopting environmental and climate friendly policies and practices. For example, recycling and waste reduction, pollution prevention, and being more energy efficient have significant cost reduction potential particularly for entrepreneurs and may help the SME save money in the long term, thereby contributing to long-term success of the business. Adopting effective environmental and climate change policies and practices also increases cooperation between the SME and the local communities, enhances customer loyalty and business reputation.

Marketplace Policies and Practices

The importance of developing and managing strategic alliance with internal and external partners is integral to SME success and to attaining a good brand and business reputation. It is no longer enough for organizations to "say the right things" they must be seen to "do the right things" in order to maintain their reputation and working relationships. The findings in Table 12.4 below show that the entrepreneurs are aware of the importance of engaging in transparent, honest, fair, and equitable practices with customers, suppliers, and business partners. The result vis a vis the SME's ability to work together with other companies/organizations to address issues raised by responsible entrepreneurship shows that more awareness is required on the importance of knowledge sharing and networking. Immense benefits can be obtained from enterprise networking, and this helps to enhance the collaborative competence of the firm, external organizations and institutions through information and technology exchange, training workshops, and research and development. Effective enterprise networking is critical to SME success and innovation.

Local Community Policies and Practices

The local community is a key stakeholder. Consequently, it is essential that SME's are proactive in positively engaging with the local community. Devising and adopting policies and business practices that help contribute to the enhancement of the

Table 12.4 Marketplace policies and practices

CSR and Responsible Entrepreneurship: Marketplace Policies and Practices (MP15–20)	Male Percentage	Female Percentage	Cumulative Percentage
MP15: Policy to ensure honesty and quality in all its contracts, dealings, and advertising; for example, fair purchasing policy, consumer protection	68%	83%	75%
MP16: Supplies clear and accurate information and labelling about products and services, including its after-sales obligations	70%	66%	68%
MP17: Ensures timely payment of suppliers' invoices	86%	89%	88%
MP18: Process to ensure effective feedback, consultation, and/or dialogue with customers, suppliers, and the other people the SME does business with	78%	86%	82%
MP19: Registers and resolves complaints from customers, suppliers, and business partners	84%	86%	85%
MP20: The SME works together with other companies/organizations to address issues raised by responsible entrepreneurship	51%	66%	58%

local communities should be at the forefront of any responsible entrepreneurial agenda. Responsible entrepreneurship involves the SME engaging in various activities that benefit the local community. These range from sporting sponsorships, educational scholarships, and cultural initiatives to initiatives that help tackle social issues such as unemployment and crime prevention. Positive community engagement can help the SME identify new niche markets, attract new customer base, identify business opportunities, build contacts with community leaders, local government authorities, and help facilitate community projects through new business partnerships (Grayson & Hodges, 2017). The findings in Table 12.5 below show that the entrepreneurs are aware of the importance of good community policies in contributing to their organization success. Engaging in a variety of activities that benefit the local community can help the entrepreneur foster knowledge sharing and networking collaborations, attract and retain employees, enhance brand and business reputation as well as increase customer recognition and loyalty. The finding shows that more female entrepreneurs were seen to engage with their local community compared to their male counterparts. There is a need for male entrepreneurs to show commitment to locally produced products or services, encourage their employees to participate more in local community activities and increase their financial support to local community activities and projects through charitable donations or sponsorship.

Table 12.5 Community policies and practices

CSR and Responsible Entrepreneurship: Community Policies and Practices (CP21–25)	Male Percentage	Female Percentage	Cumulative Percentage
CP21: Training opportunities to people from the local community through apprenticeships or work experience for the young or for disadvantaged groups	84%	91%	71%
CP22: Open dialogue with the local community on adverse, controversial, or sensitive issues that involve SME's; for example, accumulation of waste outside the business premises, vehicles obstructing roads or footpaths	81%	86%	72%
CP23: Commitment to purchase locally produced products or services	68%	80%	86%
CP24: Encourage employees to participate in local community activities; for example, providing employee time and expertise	65%	74%	67%
CP25: Giving financial support to local community activities and projects through charitable donations or sponsorship	62%	69%	81%

Company Value Policies and Practices

The Company's core values are the set of guiding principles and philosophies that help it define its place in the wider society. A company's core values are usually contained in a mission and vision statement and inextricably linked with the company's reputation. While SME's values often echo their owner's existing value systems, philosophies and ideological beliefs, the challenge is in aligning these values to the entrepreneurial context. It is important that the company value recognizes and clearly articulates how the SME's activities affects key stakeholder groups such as employees, customers, suppliers, and the local community. The findings in Table 12.6 below show an encouraging assessment of the entrepreneurs' awareness of their organization value. Majority of the SME's stated that they have clearly defined values which they communicate to their various stakeholders and they ensure that their employees are trained on the importance of the enterprise values and rules of conduct. The finding also shows that male entrepreneurs lagged behind their female counterparts in clearly defining their enterprise vales and rules of conduct.

Factor Analysis

Exploratory factor analysis was used as a reduction mechanism in order to focus on the most significant responsible entrepreneurship issues. The responsible entrepreneurship issues were grouped into five main aspects, namely; workplace and employees; environment and climate; marketplace; local community; and company value. The Principal Axis of extraction with non-orthogonal rotation (Direct Oblimin) was conducted in order to determine construct validity. The Chi-squared Test shows a significant p value at <0.001. The exploratory factor analysis yielded a

Table 12.6 Company value

CSR and Responsible Entrepreneurship: Company Value (CV26–30)	Male Percentage	Female Percentage	Cumulative Percentage
CV26: Clearly defined enterprise's values and rules of conduct	62%	80%	71%
CV27: Communicate enterprise's values to customers, business partners, suppliers, and other interested parties through sales presentations, marketing material, or informal communication	73%	83%	78%
CV28: Providing customers with information on the enterprise values and rules of conduct	84%	89%	86%
CV29: Ensure employees are aware of the enterprise's values and rules of conduct	81%	89%	85%
CV30: Training employees on the importance of the enterprise values and rules of conduct	84%	91%	88%

Four-Factor solution with Eigenvalues greater than 1. The Scree Plot in Fig. 12.1 below also shows a clear inflection at Four-Factors. The variables loaded mainly on one Factor each, and as such are distinguishable when considering the specific variables.

Table 12.7 and the Path Diagram in Fig. 12.2 below show that Factor 1 loaded more components compared to Factors 2, 3, and 4. The items retained for each Factor show that Eight components labelled CV28; EP11; EP12; EP13; EP14; EP15; WPP3; and WPP4 measuring company value, environmental and climate change, and workplace policies and practices loaded on Factor 1. In addition, Factor 2 contained Five components labelled CV29; EP9; MP17; WPP1; and WPP2 measuring company value, environmental and climate change, marketplace and workplace policies and practices. Factor 3 contains three components labelled CV30; MP18; and MP19 measuring company value and marketplace policies and practices. Finally, Factor 4 loaded three components labelled MP16; MP20; and WPP5 measuring marketplace and workplace policies and practices.

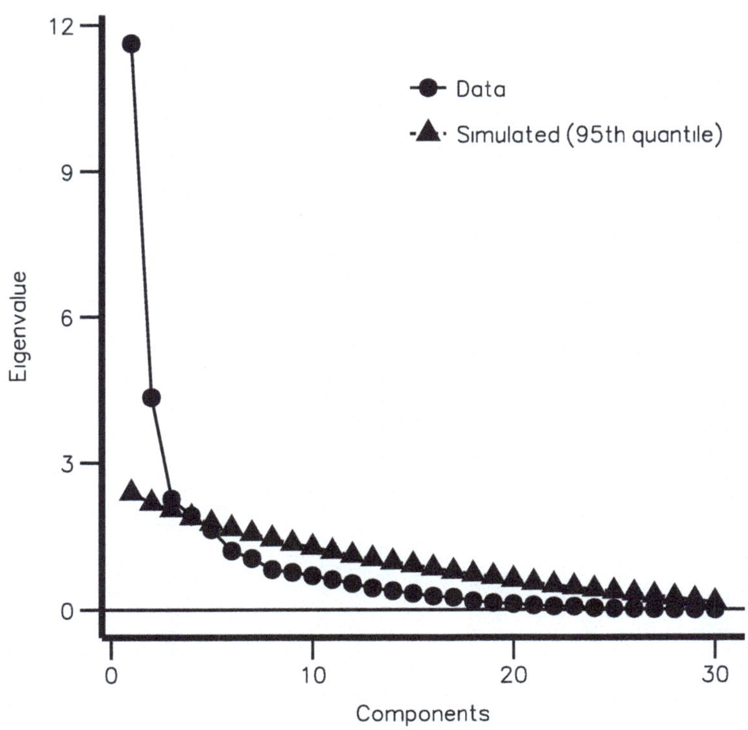

Fig. 12.1 Scree Plot

Table 12.7 Exploratory factor analysis

CSR and Responsible Entrepreneurship	Factor			
	1	2	3	4
CV28: Providing customers with information on the enterprise values and rules of conduct	0.674			
EP11: Engages in active assessment on environmental impact through recycling, reducing energy consumption, preventing pollution	0.530			
EP12: Considers potential environmental impacts when developing new products and services; for example, assessing energy usage, recyclability, or pollution generation	0.688			
EP13: Supplies clear and accurate environmental information on its products, services, and activities to customers, suppliers, and local community	0.748			
EP14: Seeks to adopt practices where the enterprise could use the sustainability of its products and services to gain an advantage over competitors; for example, recyclability of products, energy efficiency	0.613			
MP15: Policy to ensure honesty and quality in all its contracts, dealings, and advertising; for example, fair purchasing policy, consumer protection	0.756			
WPP3: Employee consultation on important issues	0.683			
WPP4: Suitable health, safety, and welfare plan for employees	0.716			
CV29: Ensure employees are aware of the enterprise's values/rules of conduct		0.661		
EP9: Reduces environmental impact by protecting the natural environment		0.786		
MP17: Ensures timely payment of suppliers' invoices		0.564		
WPP1: Developing skills and long-term careers goals by encouraging employees through training plan/performance appraisal		0.822		
WPP2: Workplace policies to ensure adequate steps are taken against discrimination in recruiting staff and in the workplace		0.967		
CV30: Training employees on enterprise values/rules of conduct			0.559	
MP18: Process to ensure effective feedback, consultation, and/or dialogue with customers, suppliers, and the other people the SME does business with			0.794	
MP19: Registers and resolve complaints from customers, suppliers, and business partners			0.834	

	Factor			
CSR and Responsible Entrepreneurship	1	2	3	4
MP16: Supplies clear and accurate information and labelling about products and services, including its after-sales obligations				0.589
MP20: The SME works together with other companies/organizations to address issues raised by responsible entrepreneurship				0.597
WPP5: Work–life balance for employees through flexible working hours or working from home				0.640

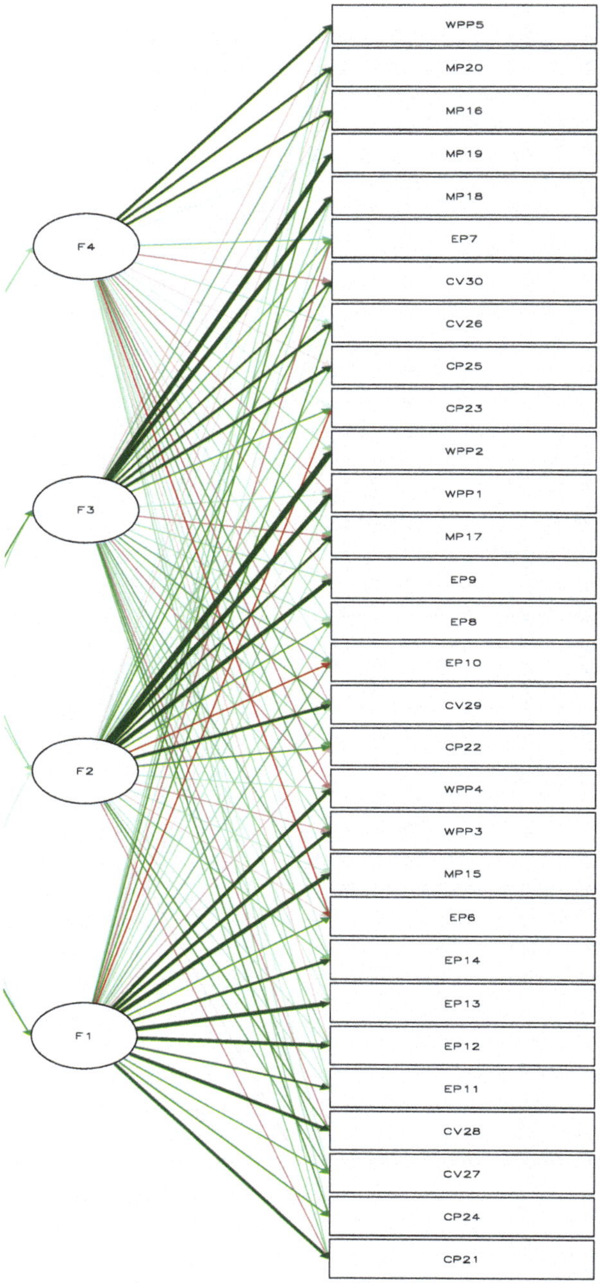

Fig. 12.2 Path diagram

Conclusion and Recommendation

This study sought to evaluate the awareness of male and female SME's entrepreneurs in Abeokuta on CSR and responsible entrepreneurship and to propose practical solutions in ensuring design and implementation of effective policies vis a vis workplace practices and employee involvement, environmental climate practices, marketplace practices, engagement with the local communities, and company values. The findings suggest a strong link between the entrepreneurs' personal moral and ethical values and their adoption of CSR practices and responsible entrepreneurial practices. The findings show that both male and female entrepreneurs would benefit from enterprise networking and knowledge sharing in further enhancing their practices. The paper makes the following recommendations:

- *Employee involvement:* Creating shared partnership with employees by building a strong internal partnership network with your employees. Effective internal partnerships result in higher trust and cooperation between the firm and its employees and promote knowledge sharing among the workforce and would in the long-term help to build a good reputation for the company. It is important to engage employees in key decision-making to make them feel valued and also help develop and empower them. Identify a clear career path for each employee and put in place training and development to enable them acquire skills needed to advance.
- *Market place:* Adopting practices and policies that foster cooperation and a good relationship with various stakeholder groups is key to ensuring a good reputation. Having a clear customer compliant policy and compliant handling procedure in place and ensuring employee and customers are aware of the procedure show a transparent approach. Paying suppliers in a timely manner also ensures goodwill and a commitment to ethical business practices.
- *Environment:* Creation of an environmental policy and mapping the firms' environmental progress. Environmental policies should look to reduce energy consumption, engage in waste reduction and recycling materials. In addition, the firm should work toward communicating its environmental impact to its employees, customers, suppliers, and the local community.
- *Local community:* Engaging in activities that align with the firms objectives as well as those that interest potential customers or address the firms' employment and training/development needs. Tackling societal issues such as unemployment, upskilling, crime watch and prevention. In addition, aligning the firms' core activities to those that are benefit and value to the local community in which the firm operates and actively encouraging employees to participate within the local community such as actively participating in local sporting events, educational program, and cultural exhibitions.
- *Company values:* Engage in dialogue with key stakeholder groups and effectively communicate the firms' values and ethos internally to employees and externally to customers, suppliers, and the local community. This can be done through the firm's website and product labelling.

Study Limitations

Future research could benefit from adopting a mixed method approach to data collection involving the use of both qualitative and quantitative research techniques. In addition, a multi-state comparison of CSR awareness is needed in order to ascertain if there are any significant correlations between socio-cultural influences and level of advancement in each State. Further research would also benefit from comparing the financial success of the SME's relating to turnover profitability and growth over time with the CSR practices adopted. Finally, the perception of employees', their commitment and length of time with the SMEs would also provide a depth of knowledge to the research topic.

References

Andersson, M. (2018). American Millennials: An ideological overview. *Honors Senior Capstone Projects, 35.*

Blomback, A., & Wigren, C. (2009). Challenging the importance of size as determinants for CSR activities. *Management of Environmental Quality: An International Journal, 20*(3), 255–270.

Bowen, H. R. (1953). *Social responsibilities of the businessman*. Harper & Row.

Braun, P. (2010). Going green: Women entrepreneurs and the environment. *International Journal of Gender and Entrepreneurship, 2*(3), 245–259.

Bryman, A., & Bell, E. (2018). *Business research methods* (3rd ed.). Oxford University Press.

Carroll, A. B. (1979). A three-dimensional conceptual model of corporate performance. *Academy of Management Review, 4*(4), 497–505.

Carroll, A. B. (1991). The pyramid of corporate social responsibility: Toward the moral management of organizational stakeholders. *Business Horizons, 34*(4), 39–48.

Davies, K. (1973). The case for and against business assumptions of social responsibilities. *Academy of Management Journal, 16*(2), 312–322.

Davis, K. (1960). Can business afford to ignore social responsibilities? *California Management Review, 2*(3), 70–76.

Dine, J. (2012). Jurisdictional arbitrage by multinational companies: A national law solution? *Journal of Human Rights and the Environment, 3*(1), 44–69.

Dodge v. Ford Motor Co. [1919]. 170 Michigan N.W. 668.

Fassin, Y. (2008). SMEs and the fallacy of formalizing CSR. *Business Ethics: A European Review, 17*(4), 364–378.

Freeman, R. E., Wicks, A. C., & Parmar, B. (2004). Stakeholder theory and the corporate objective revisited. *Organization Science*, 15(3), 364–369.

Frederick, W.C. (1960). The growing concern over business responsibility. California Management Review, 2(4), 54–61. https://doi.org/10.2307/41165405.

Friedman, M. (2002). The social responsibility of business is to increase its profits. In L. P. Hartman (Ed.), *Perspectives in business ethics* (2nd ed., pp. 260–264). The McGraw-Hill Companies.

Fry, L. W., Keim, G. D., & Meiners, R. E. (1982). Corporate contributions: Altruistic or for-profit? *Academy of Management Journal, 25*(1), 94–106.

Garriga, E. (2004). & Melé D, (2004) corporate social responsibility theories: Mapping the territory. *Journal of Business Ethics*, 53, 51–71.

Giovanna, C., De Massis, A., & Lucio, C. (2012). Corporate Social Responsibility: A survey among SME's in Bergamo. *Procedia- Social and Behavioral Sciences, 62*, 325–341.

Grayson, E. G., & Hodges, A. (2017). *Corporate social opportunity! Seven steps to make corporate social responsibility work for your business*. Routledge.

Hagerström, A., & Olivecrona, K. (1953). *Inquiries into the nature of law and morals*. Almqvist & Wiksell.

Hart, A., Lionel, H., & Green, L. (2012). *The concept of law*. Oxford University Press..

Holmes, O. W. (1897). The path of the law. *Harvard Law Review, 10*, 457–478.

Hoque, N., Abdul Rahman, A. R., Molla, R. I., & Hanifa, A. (2018). Noman, and Mohammad Zahid Hossain Bhuiyan, 'Is Corporate Social Responsibility pursuing pristine business goals for sustainable development?'. *Corporate Social Responsibility and Environmental Management, 25*(6), 1130–1142.

Hsu, J., & Cheng, M. (2012). What prompts small and medium enterprises to engage in Corporate Social Responsibility? A study from Taiwan. *Corporate Social Responsibility and Environmental Management, 19*(5), 288–305.

Jamali, D. (2009). Constraints and opportunities facing women entrepreneurs in developing countries. *Gender Management Journal, 24*(4), 232–251.

Jenkins, H. (2006). Small business champions for Corporate Social Responsibility. *Journal of Business Ethics, 67*, 241–256.

Johnson, H. L. (1971). *Business in contemporary society: Framework and issues*. Belmont: Wadsworth Publishing P50.

Jones, T. M. (1980). *Corporate social responsibility revisited, redefined*. University of California Press.

Kekes, J. (1984). Ought implies can' and two kinds of morality. *The Philosophical Quarterly, 34*(137), 459–467.

Kuhn, T. S. (2012). *The structure of scientific revolutions*. University of Chicago Press.

Kuldova, T. (2018). When elites and outlaws do philanthropy: On the limits of private vices for public benefit. *Trends in Organized Crime, 21*, 295–309.

Lepoutre, J., & Heene, A. (2006). Investigating the impact of firm size on small business social responsibility: A critical review. *Journal of Business Ethics, 67*, 257–273.

Lincoln, A. (2011). Assessing Nigerian female entrepreneur's access to finance for business startup and growth. *African Journal of Business Management.*

Lincoln, A. (2017). A consideration of how small and medium scale enterprises attempt to deal with corporate social responsibility challenges within the Nigerian environment. In S. Vertigans, S. O. Idowu, & A. S. Burlea (Eds.), *Corporate Social Responsibility in challenging times: An introduction*. Springer. https://link.springer.com/chapter/10.1007/978-3-319-52839-7_8

Lincoln, A. (2018). Nigerian entrepreneurship in the twenty-first century: Corporate social responsibility challenges and coping strategies of Nigerian male and female entrepreneurs. In D. Opoku & E. Sandberg (Eds.), *Challenges to African entrepreneurship in the 21st century* (pp. 117–146). Palgrave Macmillan. https://link.springer.com/chapter/10.1007/978-3-319-61000-9_5

Lincoln, A., Adedoyin, O., & Croad, J. (2015). Fostering Corporate Social Responsibility among Nigerian small and medium scale enterprises. In S. O. Idowu (Ed.), *Corporate social responsibility: A global dimension of key initiatives implemented by corporate entities*. Springer. https://link.springer.com/chapter/10.1007/978-3-319-21641-6_18

Logsdon, J. M., & Van Buren, H. J. (2008). Justice and large corporations: What do activist shareholders want? *Business & Society, 47*(4), 523–548.

Maitland, A. (2002, May 16). Affordable responsibility: Small suppliers. *Financial Times, 14*.

Matten, M., Crane, A., & Chapple, W. (2003). Behind the mask: Revealing the true face of corporate citizenship. *Journal of Business Ethics, 45*(1–2), 109–120.

McQuoid-Mason, S. J. (2016). When are doctors legally obliged to stop and render assistance to injured persons at road accidents? *South African Medical Journal, 106*(6), 575.

Moir, L., & Taffler, R. (2004). Does corporate philanthropy exist? Business giving to the arts in the UK. *Journal of Business Ethics, 54*(2), 149–161.

Murillo, D., & Lozano, J. (2006). SMEs and CSR: An approach to CSR in their own words. *Journal of Business Ethics, 67*, 227–240.

Norman, W., & MacDonald, C. (2004). Getting to the bottom of 'triple bottom line'. *Business Ethics Quarterly, 14*(2), 243–262.

Ortiz, A., & Kuhne, S. (2008). Implementing responsible business behavior from a Strategic Management Perspective: Developing a framework for Austrian SME's. *Journal of Business Ethics, 82*, 463–475.

Perrini, F., Russo, A., Tencati, A., & Vurro, C. (2007, July). *Going beyond a long-lasting debate: What is behind the relationship between corporate social and financial performance?* Working paper, Bocconi University, Milano.

Saunders, M., Lewis, P., & Thornhill, A. (2015). *Research methods for business students* (7th ed.). Pitman.

Schuster, R., & Rachlin, H. (1968). Indifference between punishment and free shock: Evidence for the negative law of effect 1. *Journal of the Experimental Analysis of Behavior, 11*(6), 777–786.

Shaw, B. (1988) 'A reply to Thomas Mulligan's "Critique of Milton Friedman's Essay 'The social responsibility of business to increase its profits." *Journal of Business Ethics, 7*(7), 537–543, 538.

Sweeney, L. (2007). Corporate social responsibility in Ireland: Barriers and opportunities experienced by SMEs when undertaking CSR. *Corporate Governance, 7*(4), 516–523.

UN-Habitat. (2018). *Abeokuta city context report*. Available from https://www.globalfuturecities.org/sites/default/files/2020-07/Nigeria_Abeokuta_CCR.pdf (Accessed 10/11/2022).

United Nations. (2020). *United Nations Human Settlements Programme*. Available from https://www.globalfuturecities.org/federal-republic-nigeria/cities/abeokuta#section-overview (Accessed 10/11/2022).

Weyzig, F. (2009). Political and economic arguments for corporate social responsibility: analysis and a proposition regarding the CSR agenda. *Journal of Business Ethics, 86*(4), 417–428.

Woldie, A., Leighton, P., & Adesua, A. (2008). Factors influencing small and medium enterprises (SME's) in Nigeria: An exploratory study of owner/manager and firm characteristics. *Journal of Banks and Bank System, 3*, 5–13.

Wood, D. J. (1991). Corporate social performance revisited. *Academy of Management Review, 16*(4), 691–718.

Adebimpe Lincoln holds an LLB and an LLM in Commercial Law from Cardiff University and an MBA in International Business and PhD in Entrepreneurship and Business Development from the University of South Wales. She holds a Postgraduate Diploma in Legal Practice and also holds various teaching qualifications in Higher Education. She is a Fellow of the Higher Education academy and an Associate Member of the Chartered Institute of Personnel Development. She has 19 years of experience in Higher Education. She has held positions in Cardiff University, University of South Wales and was a Senior Lecturer in Law at Cardiff Metropolitan University before taking up a position as an Assistant Professor in Saudi Arabia. Adebimpe Lincoln is a legal practitioner. She also works with University of Liverpool online Master's in Law Programmes, Kaplan Online Learning and is an Adjunct Professor with Robert Kennedy College, Switzerland. Adebimpe Lincoln has vast experience supervising research students. She has supervised a wide range of Master's level research including MSc, MBA, and LLM dissertations. She has also supervised students at the doctoral level in the area of Entrepreneurship and Corporate Governance. Her research interests lie in the area of Female Entrepreneurship, Small and Medium Sized Enterprise (SME) sector, Leadership Practices of Nigerian Entrepreneurs, Corporate Governance and Board Diversity, Sustainability and Corporate Social Responsibility among Nigerian SMEs. She has published and presented a number of articles on Leadership, CSR, Governance, and Entrepreneurship in Nigeria.

Brendhain Diamond is a graduate of the University of Western Kentucky (USA), where he holds a Bachelor of Science with double majors: Business Management and Philosophy. Brendhain holds a Master's Certificate in Problem-Based Learning from Stenden University, Leeuwarden, The Netherlands. He is also a graduate of the University of Liverpool earning a Master's of Law (LLM) with Distinction. Brendhain received the graduation awards of Student of the Year (highest grade point average) and Dissertation of the Year. Brendhain Diamond lectures Economics and Marketing in Göteborg, Sweden. The main focus has been towards management programmes within higher education for the Hospitality and Tourism Industries. Additionally, he has over 25 years of experience supervising student research projects within these programmes. His research interests lie in the area of Economic and Social Development, Sustainability, and Corporate Social Responsibility.

Dr Jane Croad is a researcher and education consultant. Before commencing her academic career, Dr Croad enjoyed a successful corporate career and held management positions with Gwalia Housing, Johnson & Johnson, and the Wicker Group. Dr Croad was a Senior Lecturer in Marketing before joining Robert Kennedy College as an Adjunct Professor. Dr Croad's research focus is on urban development; sustainable city branding; corporate social responsibility and sustainability.

Index

A
Abeokuta, 218, 221–228, 233
Accountability, 6, 22, 46, 59, 74, 117, 140, 171, 201, 220
Agenda 2030, 131–132
Albertine, 172, 173, 175, 179, 181, 183–185, 187, 188
Andean Pact, 126, 128
Antarctic Treaty, 129
Authentic Legitimacy Theory
Awareness, 28, 33, 48, 64, 78, 87–90, 92, 93, 96, 98, 99, 107, 108, 194, 218, 223, 225, 226, 228, 233, 234

B
Bad practice, 60
Banking, 81, 94–98, 112, 119
Barriers, 89–91, 93, 109, 193–206, 219
Bewley's Coffee and Tea Company, 25, 27, 31, 36
Bolivia, 73–81, 128
Business generation, 36
Business responsibilities, 13, 46, 141, 145

C
Carroll's Pyramid, 210, 211
Cerro Dominador Solar Power Plant, 130
CFAO Motors Uganda, 27–30, 38
Charitable donations, 203, 205, 215, 227
Chile, 125–136
Climate, 129–132, 222–225, 228, 229, 233
Climate-change, 30, 96, 108, 110, 115, 117, 129, 150, 155, 156, 187, 222, 225, 229
Cohesion, 131, 134
Community Mobilization, 187
Community projects, 203, 205, 227
Comprehensive and Progressive Agreement for Trans-Pacific Partnership (CPTPP), 126, 128
Conceptual approach, 103–119
Constitution, 132, 134–136, 179
Corporate accountability, 4, 7, 13–15, 51, 68
Corporate failures, 8, 59–61, 150
Corporate governance, 53, 59–68, 142, 143, 161, 197, 201
Corporate greed
Corporate ineptitude
Corporate malfeasance, 60
Corporate negative environmental impacts, 11, 12
Corporate social disclosures
Corporate social responsibility (CSR), 3, 21, 45, 73, 87, 108, 139, 171, 193, 209
Crises, 8, 49, 61, 66, 87–90, 97, 98, 110, 127, 131, 151, 224
CSR activities, 5, 10, 12, 21–39, 46, 50, 75, 88, 93–95, 97, 98, 103–109, 111–118, 140, 147, 171, 172, 174, 177, 187, 193, 195, 196, 202, 204, 205, 217, 219, 220
CSR budgets, 90, 92, 98, 115, 117, 118

U. Nwagbara et al. (eds.), *Corporate Social Responsibility Disclosure in Developing and Emerging Economies*, CSR, Sustainability, Ethics & Governance, https://doi.org/10.1007/978-3-031-61976-2

CSR initiatives, 3–7, 10, 15, 45, 46, 81, 88, 89, 92, 94–97, 103, 104, 110, 112–119, 139–141, 143, 151, 159, 194–198, 204, 205, 219, 220
CSR legal framework, 4, 7, 10–12
CSR motivation, 3, 45–54, 194, 195, 202–203, 205
CSR reporting/disclosure, 46–50, 52, 54, 73–81, 94, 95, 97, 152, 187
CSR, social, 7, 13, 47, 74, 103–119, 143, 144, 172, 176–178, 194
Customers, 28, 29, 61, 73–81, 91, 94, 96, 103, 106, 115, 205, 210, 220, 225–228, 230, 231, 233

D

Dark sky parks, 131
Database, 90, 91, 95–98
Democracy, 125–136
Democratic processes, 136
Developing countries, 3, 10, 13, 22, 25, 32, 33, 45, 46, 79–81, 89, 92, 94, 107, 140–142, 150, 151, 156–159, 162, 186, 196, 197, 218
Developing economy, 196
Development in transition, 132
Domestic companies, 93
Downstream, 173, 174, 186, 187
Duty of care, 7, 10

E

Economics, 8–11, 14, 22, 34, 37, 38, 48, 51, 59, 61, 73–80, 88, 89, 98, 103–119, 125, 126, 132, 134, 135, 140, 141, 143–146, 148–150, 153–158, 160, 161, 171, 172, 174, 177, 181, 187, 193–195, 197, 209, 211, 212, 218, 221, 222
Employees, 24, 25, 27, 49, 53, 63–68, 75, 77, 91–93, 95, 96, 103, 106, 107, 111, 113, 116–118, 147, 161, 171, 177, 187, 194, 197, 201–203, 205, 210, 214, 218, 221, 223–224, 227, 228, 230, 231, 233, 234
Employee welfare initiatives, 203
Endemic corruption, 62
Energy, 75, 87, 88, 108, 117, 129, 130, 133, 155, 222, 225, 230, 233
Entrepreneurship, 130, 218–228, 230, 231, 233
Environmental, 4, 24, 45, 73, 93, 103, 126, 139, 195, 220
Environmental degradation, 4, 132, 224
Environmental projects, 203, 205
Environments, 45–47, 139–142, 145, 147, 148, 150–156, 159–161, 193, 201, 218, 220, 223–226, 228, 230, 233
Escazú Agreement, 4–8, 10, 12, 14, 15, 23, 27, 28, 31, 35, 36, 60, 62, 64, 68, 74, 75, 80, 81, 88, 91–94, 96, 103–107, 112, 115, 117, 119, 125–127, 129, 132, 133, 171, 174–177, 181, 183, 184, 187
Estallido social, 134
Ethical theories, 23, 149, 210, 214
Ethics, 25, 50, 52, 53, 59, 106, 116, 143, 145, 197, 201, 214, 218
Ethics codes, 203
EU-Chile Advanced Framework Agreement, 129
Extractive Industries Transparency Initiative (EITI), 171, 178, 186

F

Fairness, 62, 94
Financial performance, 13, 51, 52, 64, 88–92, 94, 95, 97, 116, 152, 202
Firm performance, 152
Forestry, 93, 97, 153
Founders, 22, 26, 29, 31, 34–37, 39, 194

G

Gas flaring, 7–9, 15
Gender, 109, 118, 127, 131, 172, 175, 187, 200, 222
Global memorandum of understanding (GMOU), 4–7, 10, 11, 15
Governance, 46, 60, 74, 104, 130, 140
Governance challenges, 63
Governance framework, 60
Governance philosophy, 214
Green hydrogen, 129, 132, 133
Green hydrogen strategy, 130, 132

H

Host communities, 5–8, 10–12, 14, 15
Host Community Development Trust (HCDT), 10–12, 14, 15
Humanism, 21–26, 29, 31, 33–35, 37–39
Humanistic, morals, 23
Human rights, 4, 9, 10, 13, 61, 75–81, 106, 114, 126, 127, 129, 132, 134, 145, 155, 201, 218

I
Illegality, 150, 213
Inept system
Institutionalizing CSR, 21, 37–39
Institutional voids
Institutions, 24, 46, 61, 75, 94, 131, 154, 222
Instrumental theory, 210–212, 214
Integrating social demands theory, 213–214

L
Labour, 24, 25, 106, 117, 129, 150, 158, 181, 182
Land, 127, 133, 143, 155, 172, 175, 179, 187, 194
Latin America, 78, 80, 125–127, 130
Law-backed CSR model, 4
Legal responsibilities, 14
Legitimacy, 24, 45–47, 50–54, 105, 139, 140, 143, 144, 149–152, 159, 162, 210, 213, 215–219
Light pollution, 131
Limited views and lexicographic utility theory, 214–215
Listed companies, 92, 93, 95, 97
Local capacity building, 222
Local governments, 175, 187, 227
Loyalty, 60, 91, 94–96, 118, 157, 205, 225, 227

M
Marketing, 29, 75, 96, 174, 180, 181, 228
Market value, 92
Mercosur, 126–128
Mindset Programme, 187
Mineral Development
Morality and societal norms, 215–218
Motivations, 3, 47, 49, 62, 64–67, 77, 115, 145, 193–206, 213
Multilateralism, 125–136, 141
Multinational, 25, 27, 30, 46, 88, 92, 93, 96, 97, 119, 157, 158
Multinational corporations (MNCs), 3–5, 10, 46, 105–106, 117, 119, 140–144, 150–153, 157–162

N
National Development Plan (NDP), 173, 188
National Oil and Gas Policy, 174–186
National parks, 129
Natural resources, 117, 141, 142, 150, 151, 153, 155, 159, 172, 187, 195
Negative injunction duties, 4, 6–10, 12, 13, 15
Niger Delta, 3–12, 15, 49, 52, 150
Nigeria, 3–10, 12, 14, 15, 46, 52, 59–63, 66–68, 194, 195, 198, 204–206, 220–221, 224
Normative, 45–54, 143–145, 151, 152, 159, 161, 162

O
Ocean governance, 128, 129
Ogun State, 221
Oil industry, 3–7, 10, 15, 182
Oil spills, 7, 8, 12, 49, 184
Oltenia, 92, 97
Open-access, 90, 91, 95, 98, 222
Organisational practice, 46, 49, 54, 140, 141, 143, 149, 151
Organizational effectiveness, 65

P
Perceptions, 10, 13, 64, 94, 96, 103, 106, 113, 130, 193–206, 223, 234
Performance, 49, 60, 74, 92, 117, 146, 218
Petroleum Authority of Uganda (PAU), 172, 174, 175, 186–188
Petroleum Industry Act (PIA), 4, 7, 10–12, 14, 15
Philanthropy, 5, 21, 29, 33, 148, 174, 195, 205, 209, 212, 218
Policy implementers, 175
Political theory, 212–213
Poor leadership, 59, 66
Poor practices, 64
Positive affirmative duties, 4, 7–9, 13, 15
Practices, 3, 22, 46, 59, 73, 92, 104, 140, 172, 193, 218
Preferred Reporting Items for Systematic Reviews and Meta-Analyses (PRISMA), 90, 91, 95, 98
Private firms
Profits, 13, 21, 46, 47, 87, 98, 104, 105, 107, 110, 117, 140, 141, 143–145, 147, 148, 153, 156, 202, 203, 205, 210–212, 214–219
Psychology, 22, 36
Public firms
Public sector, 89, 90, 95, 96, 98, 133

Q
Quality of life, 22, 128, 147, 156

R

Regional integration, 126, 128, 135–136
Regulatory non-compliance, 77
Renewable energy, 93, 108, 111, 117, 129, 130, 132
Reporting, 3–15, 45, 47–49, 51, 54, 62, 73, 74, 76, 80, 81, 91, 93–97, 175, 186
Reputations, 21, 32, 37, 45, 73, 74, 95–98, 105, 106, 115–117, 125, 129, 139, 144, 148, 152, 157, 161, 202, 205, 218, 222, 225–228, 233
Research gaps, 98
Responsibility, 3, 22, 46, 59, 74, 92, 106, 140, 194, 209
Return on assets, 64, 97
Return on equity, 97
Return on investment (ROI), 151, 152
Romania, 89–98

S

Scopus, 90, 91, 95–98
Screening, 90, 91, 95–98
Self-interest, 13, 24, 36, 145, 149, 213, 215
Shareholders, 10, 46–48, 50–52, 62, 87, 140, 143, 144, 148, 149, 151, 160, 201, 210, 212, 213, 215–217, 221
Sierra Leone, 140, 142–144, 150–152, 156, 158–162
Small and medium sized enterprises (SMEs), 23, 88, 91–94, 97, 98, 103, 104, 108–111, 113–117, 119, 193–206, 218–223, 225–228, 231, 233, 234
SME values, 219, 228
Social, 4, 22, 45, 59, 73, 92, 103, 126, 139, 194, 209
Social coherence, 134–135
Social compliance, 171, 172, 174–178, 186–188
Social development, 117, 119, 154, 175, 187
Social explosion, 134
Social licence to operate, 144
Societal expectations, 216, 218
Spear Group of Companies, 25–27, 33, 34
Stakeholder engagement, 6, 46, 50, 51, 75, 140
Stakeholders, 3, 22, 45, 64, 73, 89, 103, 139, 171, 193, 211
Stakeholder theory, 51–54, 148, 210, 213–216
Sustainability, 4–6, 14, 15, 22, 26, 31, 32, 37–39, 45, 47, 50, 62, 67, 68, 73, 74, 90–93, 95–97, 99, 103, 105, 108–116, 119, 126, 127, 129–136, 139, 141, 145, 150, 153, 154, 156, 161, 162, 183, 186, 201–203, 205, 225, 230
Sustainable community development (SCD), 4–6, 10–12, 15
Sustainable development, 6, 21, 27, 48, 91, 93, 97, 110–112, 116, 119, 125–127, 129, 130, 133, 135, 139–145, 150–156, 158–162, 175, 177, 183, 187, 221–222

T

Tax regulation, 220
The firms, 23, 45–54, 73, 77, 78, 81, 108, 110, 111, 114–116, 145, 148, 195, 198, 199, 214, 219, 226, 233
The Humanism of CSR Model, 26, 37, 38
The Mercedes-Benz Group AG, 25
Toyota Motor Corporation Japan, 27
Toyota Uganda Ltd, 27, 28, 38
Transformation, 64, 66, 117, 132, 161
Transformational leadership, 59, 60, 63–68
Transparency, 4, 6, 13, 46, 47, 49, 50, 59, 61–63, 65–68, 74, 75, 79, 93, 116, 126, 140, 143, 150, 160, 161, 220
Trends, 36, 38, 51, 73–81, 88, 105, 115, 119, 144, 161

U

Uganda, 22, 25, 27–33, 35, 171–188
UN BBNJ Treaty, 128
UN Cooperation Framework UN - Chile 2023–2026, 126–127
Unilever Plc, 27, 30
Unilever Uganda Ltd, 27, 30, 31
Upstream, 11, 173, 174, 186, 187

V

Voluntary CSR practices, 4
Volunteer activities, 27, 28, 203, 205

W

Water Management Programme, 187
Weak institutions, 46, 80, 140

Z

Zero emission, 133
Zero-emission buses, 133

The manufacturer's authorised representative in the EU is Springer Nature Customer Service Centre GmbH, Europaplatz 3, 69115 Heidelberg, Germany. If you have any concerns regarding our products, please contact ProductSafety@springernature.com

Printed and bound by CPI Group (UK) Ltd, Croydon, CR0 4YY
30/09/2026
02225562-0006